Oktay Ural, Editor
CONSTRUCTION OF LOWER-COST HOUSING

Robert M. Koerner and Joseph P. Welsh
CONSTRUCTION AND GEOTECHNICAL ENGINEERING
USING SYNTHETIC FABRICS

J. Patrick Powers
CONSTRUCTION DEWATERING: A GUID
PRACTICE

Harold J. Rosen
CONSTRUCTION SPECIFICATIONS WRIT
PRINCIPLES AND PROCEDURES, Second

D1289054

Walter Podolny, Jr. and Jean M. Müller
CONSTRUCTION AND DESIGN OF PRESTRESSED
CONCRETE SEGMENTAL BRIDGES

Ben C. Gerwick, Jr. and John C. Woolery
CONSTRUCTION AND ENGINEERING MARKETING
FOR MAJOR PROJECT SERVICES

James E. Clyde
CONSTRUCTION INSPECTION: A FIELD GUIDE TO PRACTICE,
Second Edition

Julian R. Panek and John Philip Cook
CONSTRUCTION SEALANTS AND ADHESIVES, Second Edition

Courtland A. Collier and Don A. Halperin
CONSTRUCTION FUNDING: WHERE THE MONEY COMES
FROM, Second Edition

James B. Fullman
CONSTRUCTION SAFETY, SECURITY, AND LOSS PREVENTION

Harold J. Rosen
CONSTRUCTION MATERIALS FOR ARCHITECTURE

William B. Kays
CONSTRUCTION OF LININGS FOR RESERVOIRS, TANKS,
AND POLLUTION CONTROL FACILITIES, Second Edition

Walter Podolny and John B. Scalzi
CONSTRUCTION OF CABLE-STAYED BRIDGES, Second Edition

Edward J. Monahan
CONSTRUCTION OF AND ON COMPACTED FILLS

Ben C. Gerwick, Jr.
CONSTRUCTION OF OFFSHORE STRUCTURES

*Construction of
Drilled Pier Foundations*

Construction of
Drilled Pier Foundations

David M. Greer, P. E.

Geotechnical Engineer

William S. Gardner, P. E.

Executive Vice President, Woodward-Clyde Consultants

A Wiley-Interscience Publication

JOHN WILEY & SONS

New York • *Chichester* • *Brisbane* • *Toronto* • *Singapore*

Library of Congress Cataloging in Publication Data:
Greer, David M., 1908–
  Construction of drilled pier foundations.

  (Wiley series of practical construction guides,
ISSN 0271-6011)
  Includes bibliographies and index.
  1. Piling (Civil engineering)   I. Gardner, William S.
  II. Title.   III. Series.

TA780.G735 1986   624.1'58   86-11011
ISBN 0-471-82881-5

Printed in the United States of America

10 9 8 7 6 5 4 3 2

Series Preface

The Wiley Series of Practical Construction Guides provides the working constructor with up-to-date information that can help to increase the job profit margin. These guidebooks, which are scaled mainly for practice, but include the necessary theory and design, should aid a construction contractor in approaching work problems with more knowledgeable confidence. The guides should be useful also to engineers, architects, planners, specification writers, project managers, superintendents, materials and equipment manufacturers and, the source of all these callings, instructors and their students.

Construction in the United States alone will reach $250 billion a year in the early 1980s. In all nations, the business of building will continue to grow at a phenomenal rate, because the population proliferation demands new living, working, and recreational facilities. This construction will have to be more substantial, thus demanding a more professional performance from the contractor. Before science and technology had seriously affected the ideas, job plans, financing, and erection of structures, most contractors developed their know-how by field trial-and-error. Wheels, small and large, were constantly being reinvented in all sectors, because there was no interchange of knowledge. The current complexity of construction, even in more rural areas, has revealed a clear need for more proficient, professional methods and tools in both practice and learning.

Because construction is highly competitive, some practical technology is necessarily proprietary. But most practical day-to-day problems are common to the whole construction industry. These are the subjects for the Wiley Practical Construction Guides.

M. D. MORRIS, P.E.

Preface

The original book of this series, *Drilled Pier Foundations*, was published in 1972, and was out of print about five years later. The present authors, in planning a revised edition, decided that there is so much new material that two volumes are needed: *Construction of Drilled Pier Foundations* (this book) and *Design of Drilled Pier Foundations* (now in preparation).

This book was written not only for the people doing the actual construction, but also for the architect, geotechnical or structural engineer, and engineering technician, as well as the general contractor, developer, and project owner. We believe that anyone with a technical or financial connection with a construction project involving deep foundations can profit by familiarity with foundation types, construction methods, and quality control.

The writers are particularly indebted to the Professional Development Committee of Woodward-Clyde Consultants, who provided both encouragement and a major part of the cost of this project.

We also owe special thanks to Jim Lewis of Lewis Foundation Company, Charles Gupton of Dames and Moore, and Donald E. Geren of Mueser Rutledge Consultants, who reviewed the book in manuscript form and made many helpful suggestions.

The staff, contractor members, associate members, and technical affiliates of the International Association of Foundation Drilling Contractors (ADSC) were especially helpful in providing information and data, in answering questions, and in reviewing parts of the book as we wrote or revised them. We also owe thanks to the ASTM, the ASFE, and the Deep Foundations Institute for their counsel and permission to use excerpts from their material.

Finally, about the name *Drilled Shaft Foundations*: As we state in the first paragraph of the book, different people, in different areas, call the subject of our book by at least four different names. We take no position as to the correctness of any particular one. Instead, we hope that the reader will adopt the position so well presented by C. L. Dodgson (a.k.a. Lewis Carroll) in *Through the Looking Glass*, "When I use a word," Humpty Dumpty said in a rather scornful tone, "it means just what I choose it to mean— neither more nor less."

DAVID M. GREER
WILLIAM S. GARDNER

Fayetteville, Arkansas
Plymouth Meeting, Pennsylvania
July 1986

Contents

Construction of
Drilled Pier Foundations

1

State of the Art

1.1. Scope of the Book

This book, for contractors, engineers, architects, and developers, discusses the equipment and techniques involved in the construction of cast-in-place concrete foundations for which the excavation is made by an earth- or rock-boring machine. Such a foundation element, primarily used for the support of structures, is variously called a "drilled pier," a "drilled caisson," a "drilled shaft," or a "large-diameter bored pile" (British usage). The type of drilled pier or caisson is often identified by the prefixes "straight shaft," or "belled" or "underreamed," denoting respectively a constant-diameter and an enlarged-base shaft. Similarly, the term "socketed" denotes a shaft drilled into rock to form a "rock socket" which provides the primary resistance to the applied load.

In this book we shall use the terms "drilled pier" (or shaft) and "drilled and underreamed pier" or "drilled and socketed pier" for these foundation elements. An enlarged base will be referred to as an "underream" or as a "bell," interchangeably.

There are other types of large-diameter drilled holes in underground construction, in addition to the installation of foundations for structures, which are important enough to merit consideration in a book of this sort. We are therefore including, where the authors' experience indicates that the application may be of importance, discussion of some additional structural elements or construction techniques using the same drilling machines as are used for pier holes. Some of these are:

Very-large-diameter drilled holes for access shafts, missile silos, mineral exploration, and so on.

Slanted or battered holes for tieback anchors, drainage, or other uses (including batter piles).

Retaining walls and bracing for excavations (both tieback and cantilevered types).

Slurry trench construction (for seepage cutoff or for load-bearing walls.)

Landslide stabilization piers and drains.

This is a "state-of-the-art" book. The art continues to change. New techniques, new and evolving machines, and increasing skill on the part of the specialty contractors who do most of the work have led to many new capabilities in the field in the past decade. This will undoubtedly result in significant new developments and capabilities in the next decade.

The drilling equipment of today (1985) is far superior to its early predecessors. Through application of hydraulic and pneumatic systems supplemented with larger engines, refined drive trains, and improved drilling tools, the modern drilling machines can excavate materials previously termed "refusal" or "nondrillable."

Because of wide variety and general ease of mobility practically all types of drilling equipment are normally available throughout the United States and Canada. Selection of the proper size of unit should be carefully considered, however. Designation of a specific manufacturer or particular type of machine may limit or restrict competitive response from drilling contractors. The field-proven experience and reliability of the drilling contractor coupled with the expertise of the drilling superintendant and equipment operators can significantly influence the cost and time for completion of any project. These considerations should be kept in mind by designers, specification writers, and owners, as well as by the drilling contractors themselves. Before a contracting agency accepts a lowest bid, the bidder's experience and equipment should be reviewed by someone who understands the ground conditions and contractor's experience and equipment as they relate to the job.

The book does not include consideration of driven piles, drilled-in piles (augered piles), "displacement caissons," or "pressure-injected footings" (Franki piles). Nor does it cover drilled or mole-excavated tunnels. It deals only with installations in large-diameter machine-drilled (or bored) holes which have had all the soil removed (and usually, but not always, all the water).

Another related type of foundation not considered in this book is the "drilled-in caisson" (1.1, 1.2)* that has been used widely in New York City

---

* Numbers enclosed in parentheses refer to references at the end of the chapter.

and elsewhere in the United States where very heavy loads were involved and sound rock, usually at considerable depth, is available to carry the loads. In this foundation type, a heavy steel shell, with a very heavy steel cutting shoe, is driven to rock by a pile driver and is then cleaned out by blowing, jetting, or augering. A rock socket is then drilled with a churn drill or a pneumatic-percussion drill, a partial or full-length steel beam or reinforcing cage is positioned in the casing, and the entire system is concreted. This special type of deep foundation is well known and differs from the types of construction treated in this book in that the steel casing is very heavy and is driven to rock rather than being installed by boring techniques.

The term "large diameter" refers to holes larger than 15 in. (38 cm) and usually at least 24 in. (61 cm) in diameter. Some of the holes may be very much larger. Pier shafts 10 ft (3 m) in diameter and underreams 20 ft (6 m) or larger are not uncommon. An ASCE–ASA publication defines a foundation pier as one which is at least 2 ft (61 cm) in diameter; the Canadian National Building Code defines it as a shaft with a bearing surface having a minimum diameter of about 5 ft (1.5 m).

1.2. Drilled Piers: Advantages and Disadvantages

Among the important considerations for using drilled piers in preference to other types of deep foundation are economy and quality assurance (by the latter we mean conformance of the finished product to the design specifications).

In ground where soil, rock, and groundwater conditions are suitable for their use, drilled piers often cost less than other types of deep foundation per unit of load carried. Where soil and rock conditions are most favorable, the saving involved by the use of drilled piers can be very substantial, sometimes amounting to half the cost of the nearest competitive type of foundation. The potential economic advantage of drilled pier foundations comes principally from the fact that rotary drilling is a rapid and inexpensive way of excavating provided that: (1) a circular excavation, usually with its depth substantially greater than its diameter, can be adapted to the proposed structural requirements; (2) the material to be excavated can be cut readily with rotary drilling tools; (3) the soil does not tend to cave or flow into the hole; and (4) pier depths are not greater than can be achieved with locally available drilling equipment. As any one of these advantageous conditions diminishes, the cost of drilled pier construction goes up. As will be illustrated later, however, there are often ways of circumventing difficulties due to ground conditions and, even in the absence of ideal conditions, it can be shown that drilled piers often retain an effective cost advantage.

The quality assurance advantage provided by drilled piers over other types of deep foundations derives from the fact that a drilled pier hole, if it can be completed at all, can often be inspected to make sure that it has reached suitable bearing, and it provides the facility to "proof test" the bearing materials, using laboratory and/or in-situ techniques. This is not true of a driven pile, and there will be circumstances under which the long-term quality of the bearing material will be uncertain (recognizing that the driving resistance of a pile can be monitored and interpreted as a measure of its static load supporting capacity). Under some geologic circumstances (such as in solution- and shatter-prone rock), this is a definite advantage for the drilled pier foundation.

The following features, favorable and unfavorable, are among the considerations which may influence the choice between drilled piers and other types of foundation.

1.2.1.   Favorable Factors

Large Loads Are Supported by Single Piers. A single high-capacity pier per column can often replace a pile cluster or a heavy mat foundation, with significant cost savings.

Early Foundation Construction Is Facilitated. Drilled piers can often be completed before grading operations and basement excavation are undertaken, appreciably expediting overall completion of the job.

Piers Can Be Drilled through Cobbles and Small Boulders. Suitable machinery will drill through stones which would deflect piles, and so redesign is eliminated. (Redesign of more than 30% of the pile caps is not unusual for projects involving piles driven through bouldery soils or fill.)

Reinforcing and Form Construction Are Minimized or Eliminated. This produces savings when compared with conventional excavation and footing construction.

Pile Hammer Noise Is Eliminated. Most rotary-type pier-drilling machines are relatively quiet; however, machines using percussion or compressed air, for drilling through rock, may be very noisy.

Rapid Completion Is an Advantage. When surface and subsurface conditions are suitable, the rapid completion of piers not only can expedite the foundation phase of the project, but also can expedite subsequent operations because of minimal interference with other construction operations. (This advantage is subject to the effects of weather and of unexpected ground conditions. As a result pile foundations, though nominally more costly, may be a better choice for some jobs.)

Vibration, Heave, and Ground Displacement Are Eliminated. This is a distinct advantage when nearby structures or machinery must not be disturbed. When displacement piles are driven into saturated firm-to-stiff clays, ground heave and lateral displacement of previously driven piles may become a major problem.

Uplift Resistance Is Easily Provided. Such resistance is readily developed in ground where underreams (bells) can be formed or where the sides of the shaft in the bearing stratum have enough contact to assure the development of shearing resistance between shaft and bearing material.

Uplift or Downdrag Can Be Reduced. Uplift loads due to swelling soil, as well as downdrag loads due to settling soil ("negative skin friction"), can be avoided or minimized more readily than with most other types of deep foundation because drilled piers can be double-cased easily.

1.2.2. Unfavorable Factors

Operations Are Affected by Bad Weather. Wet weather interferes with pier drilling and concreting to a much greater extent than with pile driving.

Unfavorable Soil Conditions May Interfere. Such conditions, which may be unexpected, may introduce major difficulties and delays in drilled pier construction. This is true, of course, of any foundation system, but drilled piers are (in the authors' opinion) more sensitive to this element than are competitive types of deep foundation.

More Complete Soil Exploration Is Needed. For drilled pier applications the investigation needs to be (again, in the authors' opinion) more thorough than for most other types of deep foundation.

Building Codes and Governmental Regulations May Be Unfavorable. In some cases these are so formulated that the full load capacity cannot be used, or they impose construction restrictions that increase drilled pier costs materially.

Inspection and Technical Supervision Are Critical. These are more critical for pier drilling, and especially for the concreting operation, than for most other deep-foundation systems. (Cast-in-place concrete piles also are critical in this respect.)

Ground Loss May Be Substantial. Excess excavation and consequent settlement of adjacent structures are possible under some circumstances, and their prevention requires careful attention in both design and construction.

Surface Subsidence May Occur. Uncontrolled dewatering to facilitate pier construction may induce subsoil consolidation under the increase in effective stress accompanying groundwater drawdown.

These characteristics, and their importance, are discussed at more length at appropriate places in the chapters that follow.

1.3. Development of Large-Hole Drilling Machinery

The use of deep hand-excavated piers for foundations is probably as old as civilization. Yet up to about 1945 the holes for pier foundations were still, for the most part, hand-excavated. The two most common procedures in the United States were those described in the textbooks under the subjects "Gow caissons" and "Chicago caissons." Undoubtedly a few machines for excavating pier shafts were improvised in areas where the soils were particularly suited to machine drilling, some of them going back to the days of steam power.

The earliest record the authors have found of the use of machine-drilled piers for foundations comes from San Antonio, Texas. Mr. Willard Simpson, Sr., a consulting engineer, recognized early in the 1920s that buildings in that area could be protected from the very destructive effects of soil swelling and shrinkage only if all foundations were supported at levels below the zone of seasonal moisture variations [which in San Antonio might be 25 ft (7.6 m) or more deep]. After experimenting with a few pier foundations excavated manually with the long-handled shovels that utility companies then used for making pole holes, Mr. Simpson and Mr. Ed Duderstadt, a local contractor, adapted a well-drilling machine operated by four men pushing a capstan bar around a circular track (a machine illustrated in Dempster Mill Manufacturing Company's 1895 catalog). This worked so well in San Antonio soils that the contractor was soon operating six or eight of these machines, horse-powered instead of man-powered (Fig. 1.1). In

FIGURE 1.1. An early horse-powered bucket-auger well-drilling machine (Dempster Mill Mfg. Co., 1906 catalog).

about 1938 one of these machines was mechanized by combining it with the power unit of a steam shovel. At least one of the horse-powered machines was still in use in 1951 (1.3).

Mr. Eric P. Johnsen of Abbott-Merkt and Company relates that drilled piers with 30° bells were used under the Continental Can Company Building in the (then) Union Pacific Industrial Area in Los Angeles in 1928. The drilling machine was mounted on a truck; further details are lacking.

During the 1920s the Gow Company of New York City built and used a bucket-type auger machine, electrically powered and mounted on the turntable frame of the crawler tractor of a crane. The drill bucket, about 4 ft in diameter, was equipped with adjustable "reaming knives" that enlarged the shaft hole to diameters up to at least 7 ft (2.1 m). Bells with 30° sides (angle measured from the vertical) were formed at the bottom by hand labor, using air spades for digging and electric winches for hoisting. One pier, extending to a depth of about 120 ft (37 m), was completed (including concreting) in about 10 hr (1.4, 1.5).

The development of efficient, reliable, and commercially available machines of this sort for widespread general use did not come until after World War II. During the war years, the pressure for rapid construction of many light buildings for the armed services resulted in widespread use of the truck-mounted auger machines which had been in use for a few years by the utility companies. These "pole-hole machines" were used to drill uncased holes for small pier foundations, usually shallow and not more than 18 in. (46 cm) in diameter. The work was scheduled so that the concrete truck followed immediately after the drilling machine. No forming was necessary, and the work went very rapidly. The result of the demand was the development of many small pier-drilling contractors, the most enterprising of whom started seeking other and larger work and devising their own drilling tools and, in some cases, drilling machines. In particular, tools were devised for making an expanded base (an underream or bell) without enlarging the shaft, so that larger column loads could be carried without wasting concrete on unnecessarily large-diameter shafts. Most of the machines operating in 1946 were custom-made, each model being built to match the experience of the builder.

The greatest impetus to this development occurred in two states: Texas and California. Because of differences in geology and in the soils and the way they handled, as well as differences in structural requirements, the development of "drilled pier machines," as they soon came to be called, diverged in these two areas. The result was the eventual evolvement and production of two substantially different kinds of auger-type pier-drilling machines: the "bucket-auger" machine in California (Fig. 1.2) and the open-helix-type auger machine in Texas (Fig. 1.3). (The latter should not be confused with the much smaller "continuous-flight" auger machine that is used in soil exploration and in predrilling holes for piles.) Each type of machine had its own advantages under certain favorable circumstances, and its own following among contractors.

FIGURE 1.2. CALWELD Bucket Auger No. 200-B with 30-in.-diameter drilling bucket and a 30-ft telescoping kelly.

As these commercially available machines evolved, improvements were made in several directions, responding to the needs of the contractors who used them.

1. Larger machines were designed, leading each manufacturer to develop a series of models of different sizes and capabilities.

2. The larger machines were designed to apply higher torque to the cutting tools, enabling the machines to drill harder rock, to turn larger-diameter soil-drilling tools, and to drill deeper holes than had been possible before.

3. Telescoping "kellys" (a square or splined shaft that can slide up and down through the rotary driving head while being rotated) were developed and increased in length and cross section, so that deeper holes could be drilled without coupling and uncoupling the drill stem.

4. Some light models which had initially depended on the dead weight of kelly and auger for downward force on the cutting edges were eventually equipped with hydraulic "pulldown" or "crowd" mechanisms to produce faster penetration into hard formations.

5. Speed in setting up was greatly increased by the addition of hydraulic raising and lowering of the mast, hydraulic outrigger jacks for leveling, and in-and-out slide and rotating base for positioning the auger over the exact point for boring.

6. Crane mountings were devised for some of the larger machines (instead of the usual truck mounting), allowing the handling of longer and heavier kellys, as well as larger augers and underreaming (belling) tools, and the placing and pulling of longer and heavier casing sections (Fig. 1.4).

7. Cutting bits and core barrels for drilling rock were greatly improved, so that the higher-torque machines could drill further into harder rock than before, still using their auger-type tools or specially adapted rock-drilling attachments.

FIGURE 1.3.   Watson Model 2000 rig mounted on a 6 × 4 truck.

FIGURE 1.4.   Atlantic CLLDH "Super Duty" crane-mounted rig working on pier holes for slide stabilization on the Seattle Freeway.

8. Oil-well-drilling tools and techniques were adapted to large-diameter hole drilling (Fig. 1.5). For example, drilling mud was used to prevent the caving of water-bearing formations, permitting rapid drilling of the shaft hole (to be cased off later) down to the bearing or belling stratum.

9. New drilling tools have been devised; larger truck-type carriers have been designed and put on the market; down-hole and raise-bit drilling systems and equipment, for special ground conditions, are now available. (Figure 1.6 shows one of the larger truck-type carriers.)

1.4.   Development of the Drilled Pier

In the chapters that follow, references will be made to four types of pier, which, while similar in construction technique, differ in the way in which they are assumed, for design purposes, to transfer the foundation load to the earth. These types are illustrated in Fig. 1.7.

Straight-shaft end-bearing piers develop their primary support from end-bearing on strong soil, glacial till ("hardpan"), or rock. The overlying soil may be assumed to contribute nothing to the support of the load imposed on the pier.

Straight-shaft "friction" piers develop primary support from the materials in contact with the circumferential area of the pier, although some end-bearing is mobilized. Thus piers penetrate far enough into an assigned bearing stratum to develop the required capacity primarily by sidewall shear.

Combination straight-shaft friction and end-bearing piers are of the same construction as the two just described, but both sidewall shear and end-bearing are assigned a role in carrying the design load. When carried into rock, this pier may be referred to as a "socketed pier" or a "drilled pier with rock socket."

Belled or underreamed piers are piers with an enlarged base. In U.S. practice, the design load of these piers is usually assumed to be supported in end-bearing, although the friction component is typically incorporated in the United Kingdom. Results of load tests on piers validate the latter approach if the stiffness of the materials in contact with the shaft and of those providing end bearing are not greatly dissimilar.

FIGURE 1.5.   Modified oil-well-drilling rig for drilling a 10-ft-diameter hole more than 5000 ft deep. (Photo courtesy of Loffland Brothers Co., Tulsa, Oklahoma.)

FIGURE 1.6. Pettibone Carrier with Watson Model 3000 rig.

In the decade following World War II, in areas where ground conditions were especially suitable for the use of rotary pier-drilling machines, the economy of drilled piers over all other types of deep-foundation construction was immediately evident, and they quickly became the prevalent type. In many cities—for example, Houston, Denver, San Antonio—pile foundations continued to be used only in exceptional circumstances, usually in locations involving caving soils and great depths to suitable bearing strata.

As the use of the drilled pier became common, some designers and some contractors lost sight of the limits of the soil, groundwater, and rock conditions which had led to the advantages of drilled piers. The resulting attempts to use them where the geologic formations were not suitable for the drilling machines, tools, and techniques then available sometimes led to unsatisfactory construction performance in the form of delays, "extras," and occasional abandonment of contracts, with the job being redesigned for piling or some other form of foundation. The difficulties developed most commonly under the following circumstances: in caving soils (especially cohesionless soils below the water table), where casing had to be used, and the available machines were not suitable for placing or pulling the necessary sizes of casing; in bouldery soils, where sometimes stones were encountered which could neither be drilled nor removed intact; and in soil-rock areas, where drilling was stopped by ledges or pinnacles which had to be broken up by jackhammer before the pier hole could be completed to the planned bearing level.

These difficulties, in turn, led to the development of specialized tools, new machines, and better drilling techniques (such as the use of drilling slurries), thus expanding the range of economical application of drilled piers. As a result, drilled piers are now used in many areas and on many jobs where a few years ago their use would have been uneconomical or actually unfeasible. Where water-bearing formations have to be penetrated and cased off, the larger and more powerful machines can often enable the contractor to drill-in the casing ahead of the excavation, or to drill the

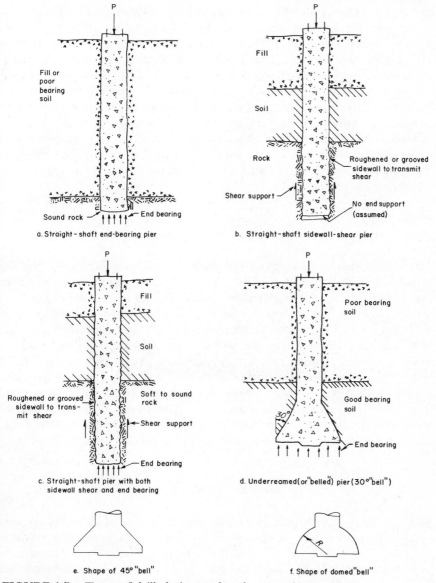

FIGURE 1.7. Types of drilled piers and underream shapes. These are "nominal" shapes. Actual shapes of some "bells" may be somewhat different. See Fig. 1.8.

hole quickly and place the casing before substantial caving has occurred. Most modern pier-drilling machines also can be adapted for pulling the casing smoothly and at a controlled rate when concrete is being placed. With the larger machines, casings can also be drilled into rock to seal off water inflow, and many contractors' experience in sealing off casing-rock contacts has led to skill and speed in this operation that were generally lacking a few years ago. The accumulation of specialized experience on the part of contractors has led to their being able to handle, sometimes as a routine operation, difficulties which would once have slowed the job greatly.

Since the development of a nationwide specialized drilled pier contracting industry in the 1960s, there has been a substantial buildup of experienced contractors and skilled machine operators. The drilling of large-diameter holes in the earth requires a very special skill, and to do it without trouble or delay under potentially troublesome conditions makes the difference between profit and loss for the contractor. It also may mean a job finished on time and without controversy, claims, and confusion as far as the owner and his agents are concerned. A substantial part of the "art" of drilled pier foundation construction lies in the personal skill of the drilling machine operator and the experience and ingenuity of the operator and the drilling contractor's superintendent.

1.5. Drilled Piers Now in Widespread Use

The local occurrence of clearly favorable ground conditions, plus the development of machinery, techniques, and skills for coping with the difficulties that arise under less favorable conditions, have led to the use—or at least the trial—of machine-drilled piers in all parts of the world. In many geologic settings they have become the dominant type of foundation. At the time of this writing (1985) there are pier-drilling contractors, either available locally or on call from a city not too far away, in any part of the United States. Drilled piers have been adopted with enthusiasm in England, Australia, Canada, and West Germany, where there is extensive current research into their bearing capacity, and their use has also spread rapidly through Europe and Latin America. Drilling machine manufacturers in the United States and other countries are now exporting their products to all continents.

1.6. Batter Piers and Anchor Piers or Tiebacks

With the early pier-drilling machines, sloping holes could be drilled only by blocking up one side (or end) of the truck on which the drilling machine was mounted. This usually limited the slope to not more than 1 horizontal to 4 to 6 vertical. Many later machines, however, have been conveniently arranged to drill with the mast and drilling assembly tilted back over the

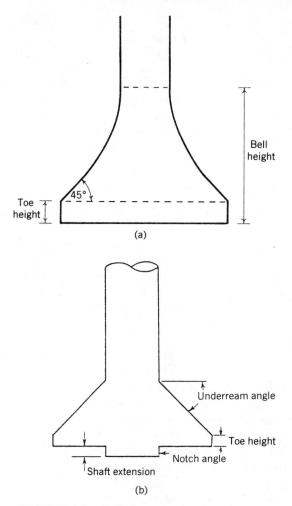

FIGURE 1.8.   Bell shapes and nomenclature.

cab (in the case of truck-mounted machines), or forward (in the case of crane-mounted rigs). As a result, where suitable equipment is available, battered piers and tieback holes can now be drilled with conventional auger machines. Batters flatter than 1 horizontal to 3 vertical, however, are difficult to keep in alignment, and are limited in economical application to the most favorable ground conditions. Pier holes with any batter at all are more expensive than vertical ones. This places a limit on their economical use, rather than on what can be done if necessary. A further limitation that must be kept in mind is that, even with ground conditions that are ideal for forming underreams in vertical pier holes, an expanded base is much more difficult to form at the bottom of a sloping hole, and will be much more expensive and less certain as to ultimate shape, dimensions, and the completeness and continuity of concreting.

Smaller-diameter holes for anchor piles or tiebacks, however, are readily drilled with specialized drilling machines. With suitable equipment they can be drilled at practically any angle, including slopes upward from the drilling point (as needed for drainage installations, for example). Because the dimensions and shapes of enlarged bases (or ends) for tieback anchor holes are not as critical as are those of the underreamed bases for piers, the disadvantage of uncertainty of base shape and dimensions mentioned above is here minimized.

1.7.   Drilled Piers in Caving Ground: Some Special Techniques

Except for rock, the most common impediment to large-hole drilling is the need to pass through (and case off) water-bearing formations. In the early years of the industry, this usually was not attempted. As the uses of large-diameter drilled holes spread—both geographically and geologically—many jobs were found to be feasible with the aid of conventional dewatering (by well points, deep wells, etc.). In these applications, the speed of completion of drilled holes, as compared with conventional excavation and forming, greatly reduced the time that dewatering had to be maintained, thus effecting an additional substantial saving. The use of minimal dewatering installations often permits pier holes to be drilled and promptly filled with concrete in ground where much more elaborate dewatering installations would have been required if more conventional excavation and construction techniques had been used.

Well points have been employed for temporary stabilization of fine-to-medium sand where capillary forces in the dewatered but moist sand were sufficient to keep the hole from caving for a few minutes. Deep wells have been used successfully for stabilizing clay with sand and gravel layers which otherwise would have produced very troublesome sloughing and caving; and in fissured clays where the deep wells, pumped down to a level 10 ft (3 m) below the bottoms of the footings, allowed 8-ft-diameter (2.4 m) underreams to be cut and concreted in the dry and without caving. Dewatering techniques thus can be used occasionally to convert a potentially troublesome operation into a near-routine one.

The use of drilling mud to prevent caving, a technique adapted from oil well practice, has been used successfully to extend the range of application of drilled piers into ground conditions that would otherwise not have permitted either the drilling of an open hole or the economical use of casing alone. The technique is often not understood by people who have not had experience with it. However, it has an important potential for economy when conditions are otherwise suitable for carrying the foundation loads on drilled piers (though unfavorable for their construction by ordinary means) and, for one reason or another, unsuitable for the installation of other types of deep foundation.

An example is an addition to a paper mill, involving foundations to support heavy loads, in a location where no disturbance to existing machinery could be tolerated and where 20 or more feet (6 m) of loose water-bearing sand overlie the bearing stratum of stiff clay. The pier shafts were drilled down to the clay in mud-filled holes and the casing was then inserted and sealed into the clay. The mud was removed from the casing, and the underream was then cut in a dry hole with the usual underreaming tool. The pier was completed without difficulty and (typically) the casing was removed and re-used. All of this was accomplished without vibration, whereas pile driving would almost certainly have disturbed the adjacent machinery in the plant.

Mud drilling techniques also permit a shaft to be drilled through caving soils and completed without the use of casing. There have been many instances of shafts *and underreams* being drilled entirely in mud-filled holes, the concreting of the completed bell and shaft being accomplished entirely by displacement of the mud (slurry), from the bottom up, with pumped-in or tremied concrete.

Another important innovation in handling the problem of casing off water-bearing sand or gravel is to install the casing with a "vibrodriver," which vibrates the casing through the troublesome stratum and seats it into an underlying impervious stratum. This method offers the advantages of speed; easy, effective seating of the casing into a relatively impervious stratum; and maximum lateral support for the pier.

References

1.1   White, R. E., Development in Large-diameter Piles or Caissons in the United States, *Proc. 3rd Asian Reg. Conf. SM & FE,* * Haifa, Israel*, pp. 79–85, 1967.

1.2   White, R. E., Heavy Foundations Drilled into Rock, *Civil Eng.*, **13**(1), 19–32 (January 1943).

1.3   Old Dobbin Comes Back, *San Antonio Light*, December 3, 1951.

1.4   Christie, H. A., Boring Machine Digs Wells for Concrete Piers, *ENR*,† July 28, 1932, pp. 105–106.

1.5   Beretta, J. W., Drilled, Reamed Holes for Bridge Foundation Piles, *ENR*,† February 11, 1937, pp. 217–218.

---

* *SM & FE = Soil Mechanics & Foundation Engineering.*
† *ENR = Engineering News Record.*

2

Construction Equipment and Tools

2.1. Drilling Machines

2.1.1. General: Types, Ratings, and Capacities; Bases for Choice

Commercially produced drilling rigs of sufficient size and capacity to drill pier shafts come in a wide variety of mountings and driving arrangements. Mountings are usually truck, crane, crawler, or skid—see Figs. 2.1 to 2.7. Driving arrangements usually fall into one of three classes: the kelly bar driven by a mechanically geared rotary table (Figs. 2.1–2.3 and 2.5); the kelly driven by a yoke turned by a ring-gear (Figs. 1.2 and 2.6); or the hydraulic drive, with the hydraulic motor either mounted at the turntable (Fig. 2.2) or mounted on top of and moving with the drill steam. Choice of these details is a matter of contractor's preference, based on experience and suitability of the machine to job and ground conditions in the areas where it will be used, and on local availability. However, the choice of size (or capacity) of the drilling machines to be used on a job may become the concern of the owner, the architect, and the structural engineer, as well as that of the contractor. The authors have encountered many instances where the successful bidder on a drilled pier contract did not have suitable or heavy enough equipment for the job. The result was invariably controversy, claims for "extras," or delayed completion—and very often all three. Suitability of the contractor's equipment for the job at hand should be established before a contract is let.

Drilling machine ratings are presented in the manufacturers' catalogs and technical data sheets are usually expressed as "maximum hole diameter,"

FIGURE 2.1. Atlantic LLDH-100 with mast down. This carrier-mounted rig is rated as being capable of drilling shafts up to 10 ft (3 m) in diameter, to depths of 120 ft (37 m).

"maximum depth," and "maximum torque" at some particular rpm (see Appendix A). The "maximum" data given are usually dictated by dimensional, strength, and power limitations. The manufacturer's intent is not to recommend that the machine be used regularly for work at these ratings, but rather to warn the user to exceed them only occasionally, under favorable drilling conditions and with additional operator caution. It may be economical for a contractor to use a drilling machine, carefully, for diameters and depths beyond its designated rating, rather than bringing in a larger and different machine for a minor amount of work. However, a drilling machine working steadily near its upper limit of capacity in any category is inefficient and liable to mechanical failure.

The curves in Fig. 2.4 show how drilling costs *per foot of hole* in a specific soil condition vary with hole diameter for a given machine, and how changing from the wrong size of drilling machine to one of more suitable capacity can reduce costs. (A similar diagram showing cost per cubic yard of soil removed would show lower minimum unit costs for successively larger machines.)

For auger drilling rock, glacial till (hardpan), or very hard soil, the available torque at auger cutting speeds and the downward force become

FIGURE 2.2. Texoma Taurus Holedigger rated at a maximum hole depth of 100 ft (30.5 m), crowd pressure 51,200 lbs; spinoff speed 281 rpm. (Photo courtesy of Reedrill, Inc., Sherman, Texas.)

major criteria of suitability. A contract involving hard drilling should never be let out unless it is certain that the contractor has equipment that is sufficiently powerful to work within its economical range under actual job conditions. In the Philadelphia area, for example, it is common practice to drill 36- to 60-in.-diameter (1.5 m) holes through overburden and very dense decomposed mica gneiss/schist. The underlying relatively unaltered rock is currently considered capable of supporting unit loads of 40 to 60 tsf (or kgf/cm^2). In order to auger drill such holes, a machine with an available minimum torque of at least 100,000 lb-ft (14,000 kgf-m) is required

FIGURE 2.3. CALWELD Model 200-CH crane attachment. Torque output of this machine is reported by the manufacturer at 360,000 lb-ft (50,000 kgf-m). (Photo courtesy of Caisson Corp., Chicago.)

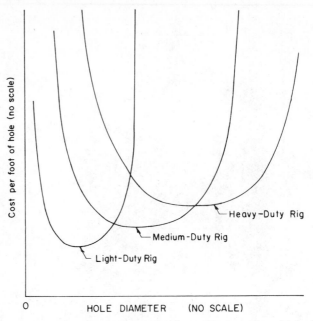

FIGURE 2.4.   Curves showing how cost per foot of hole varies with hole diameter, for various sizes of rig (not to scale).

where drilling experience has shown that lighter auger rigs cannot efficiently drill holes of this size to the required depth in these formations. Drilling through layers of hard rock within the soft rock, as well as sockets through hard rock, are facilitated by changing to rock augers and/or core barrels.

In soils, hardpan, or soft or weathered rock, the use of an auger with a drilling machine of suitable torque capacity also has proved much more economical than the use of core barrels or other tools with a lighter rig. Auger drilling machines with a nominal torque rating of 360,000 lb-ft (50,000 kgf-m) and higher are available in some cities (Fig. 2.3) (2.1, 2.2, 2.3).

The torque criterion mentioned above applies, of course, only to a machine using an auger-type tool; machines using core barrels, roller bits, or down-hole chopping bits will not require as much torque—but will usually not make holes as fast as an auger machine *of suitable capacity and suitable design* in hard soil or soft rock formations.

2.1.2.   Bucket and Auger Machines

In Appendix A, auger drilling rigs from some of the major suppliers of this type of equipment in the United States have been tabulated, with some dimensional data and rated maximum capacities (manufacturers' data). These have been separated into heavy duty, medium duty, and light duty using the manufacturers' recommended maximum hole diameter and depth as criteria.

2.1.3. *Circulating Water, Mud, or Air Machines*

Because pier foundations drilled into rock have proven economical in many regions and are now widely used, many drilling contractors are equipped with machines and tools for drilling pier holes into soft or hard rock, in diameters as large as 15 ft (4.5 m). Very large-diameter holes in rock require the use of heavy drilling machinery, such as that used for drilling mine shafts. Holes for drilled pier foundations, 6 to 8 ft (180 to 240 cm) in diameter, can be drilled with truck-mounted or crane-mounted drilling machines (Figs. 2.5 and 2.7), using appropriate tools and air or water circulation. Where necessary to develop load-bearing capacity by sidewall friction (as in soft or weathered rock), such holes can go many feet (hundreds if necessary) into rock.

Any of the large auger machines illustrated or listed in Appendix A can be converted into machines for drilling with circulating fluid (water, mud, or air) by the addition of separate equipment for circulating water or mud, or blowing air, with suitable swivel connections to the kelly. However, some manufacturers can furnish auger machines specially designed for

FIGURE 2.5.   Watson Crane-mounted Model 500-CA drilling an 8-ft-diameter (2.4 m) hole in stiff clay soil. Note the distance to which clods have been thrown during earlier trips.

FIGURE 2.6.   CALWELD Bucket Auger Model 175-C dumping spoil.

conversion to circulation drilling; such machines are more cost effective for production drilling than field-converted units.

2.2.   Drilling Machine Mounts

2.2.1.   Truck-Mounted Machines

By far the largest number of pier-drilling machines in the United States are truck mounted. Their chief—and very important—advantage over other types is that of mobility: the ability to move at highway speeds between sites and, under favorable circumstances, to maneuver easily from hole to hole within a site. Most of them are equipped also with sliding and rotating mounts, and the masts of many models can be tilted to drill battered holes— in some cases past the horizontal position. These ready adjustments make positioning of the auger over the hole location very fast and easy. Figures

2.1, 2.2, and 2.7 illustrate some of the truck-mounted rigs available. For the larger machines, specially designed trucks called "carriers" are available. One of these is illustrated in Fig. 1.6.

2.2.2. *Machines for Crane Mounting*

Rigs for crane mounting must be brought to the site by heavy-equipment trailers and are therefore less mobile and less adaptable to small jobs than truck-mounted machines. Many of these machines are furnished as "crane attachments," to be mounted on a crane of the contractor's selection.

Crane-mounted rigs are designed to handle large and heavy augers and buckets, and, in general, can drill deeper and larger-diameter holes than rigs with other types of mounting. Their driving machinery is generally higher off the ground than that of truck-mounted machines, and consequently

FIGURE 2.7.   CALWELD Model 175-C rig arranged for reverse-circulation drilling, with 6-in. R.C. swivel, 23-ft (7 m) kelly, drag bit, with water and air hoses.

they can handle taller augers and underreamers. The higher lifting capacity of the crane also allows handling of larger and heavier casing sections and reinforcing cages without the necessity of having a separate crane on the site for that purpose.

In some areas, many of the cranes available are themselves truck mounted, but more often they are carried on tracks and are therefore readily maneuverable over soft ground.

Care is required in selecting a crane for particular rig, or in adapting a crane to mount a rig, to see that no part of the crane is overstressed by application of forces that were not taken into account in the design of the crane or its boom. This requires special attention to the stresses that can be set up at the points of attachment of rig to crane. Some crane-mounted rigs have the entire rig weight suspended from the crane; others are so mounted that their weight is carried near the base of the crane boom and on the cab.

Two large crane-mounted rigs are illustrated in Figs. 1.4 and 2.3.

2.2.3. Crawler-Mounted Rigs

Crawler-mounted rigs also have to be brought to a site by heavy-equipment trailers and are therefore less mobile than truck-mounted equipment. Their on-site mobility is excellent and because of the higher mounting they can handle taller augers and underreaming tools than most truck-mounted rigs.

2.2.4. Special Rig Mounts

Specially mounted rigs for drilling pier holes in restricted spaces, areas with limited overhead, or on slopes are available in many cities. Rigs mounted on skids or other fixed-based platforms are available for use on barges and marine platforms.

2.2.5. Rodless Reverse-Circulation Drill

This drilling machine operates down-hole, suspended from a crane or other suitable support. It can be used in limited-headroom locations where there is not enough room for crane, kelly bar, and so on. The machine is made in Japan; see Appendix A for sizes, drilling characteristics, and source of the machine in the United States. A schematic drawing showing the machine with its reverse-circulation system is shown in Fig. 2.9.

2.3. Drilling and Auxiliary Tools; Hole Cutters

2.3.1. Drilling Buckets

The bucket-type drill, shown in Fig. 2.6, was used for almost all pier drilling in California for many years. The key to its success was the ingenious driving mechanism for turning the kelly: a ring-gear, large enough for the drilling

FIGURE 2.8.   Watson Model 3000 rig on crawler mount.

bucket to be lifted out of the hole and up through the gear, so that it could be swung aside and emptied into a spoil pile or into a truck. This drive was the basis for the Calweld bucket auger machine (Figs. 1.2 and 2.6).

The efficiency of the "flight auger" (see Subsection 2.3.3), which could be emptied by spinning it as it came out of the hole; the development of machines such as the front-end loader for picking up the spoils; and the demand for very large-diameter holes—all these trends have led to the development of machines which can be used with either bucket or auger.

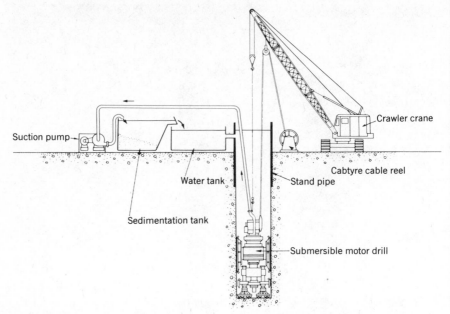

FIGURE 2.9.   Tone rodless reverse-circulation drill (Tone Boring Co., Ltd., Tokyo, Japan).

Drill buckets are still used in preference to flight augers under many circumstances, as they are often more efficient in soft soils or running sands, and the bucket is a standard tool for supplementing the flight auger in circumstances where the cuttings are too fluid or too loose to be brought out efficiently with an open helix.

Some drilling buckets are so designed that they can be used also for bailing water.

2.3.2.   Hole Reamers for Drilling Buckets

When large-diameter holes are drilled with machines using a ring-gear and yoke driving mechanism (Calweld, Earthdrill, Gus Pech) and a drilling bucket (rather than an open-helix auger), the diameter of the drilling bucket is limited by the inside diameter of the ring-gear through which the bucket must be lifted. The hole diameter, however, can be enlarged by using a reaming tool which is attached to the bucket. The reamer cuts out the full diameter of the hole and brings the cuttings into the drill bucket, which is used to bring out and dump the spoil. Holes of very large diameter can be drilled in this manner (Fig. 2.10). Some manufacturers' catalogs list reamer diameters up to 20 ft (6.1 m).

2.3.3.   Open-Helix Earth Augers ("Flight Augers")

Most drilled pier shafts through soil or soft rock are now drilled with the open-helix auger. This tool may be equipped with a knife-blade cutting edge for use in most homogeneous soils, or with hard-surfaced teeth for cutting stiff or hard soils, stony soils, or soft to moderately hard rock. These augers are available in diameters up to 10 ft (3 m) or more. Two auger bits, one for very large holes, are shown in Fig. 2.11.

2.3.4. Rock Augers

For ground conditions where penetration by a simple open-helix earth auger is stopped or slowed by hardpan, or by soft or layered rock, auger bits especially designed for rock drilling have been developed. Variations of these include the simple cast-steel auger bit with a pilot "stinger" illustrated in Fig. 2.12 and a bit set with staggered rows of hard-metal teeth (Fig. 2.13).

FIGURE 2.10. Twelve-foot (3.65 m) hole with rings and lagging sets installed; drilled with CALWELD Model 150-CH crane attachment unit, using bucket with reamers. Bucket is full, ready to come out and dump spoil.

(b)

FIGURE 2.11. (a) 10-foot diameter auger for soil hardpan. (Photo courtesy of Harrison Drilling Company, Tulsa, Oklahoma.) (b) Flight auger with hard-metal-cutting teeth, for very stiff soil or hardpan.

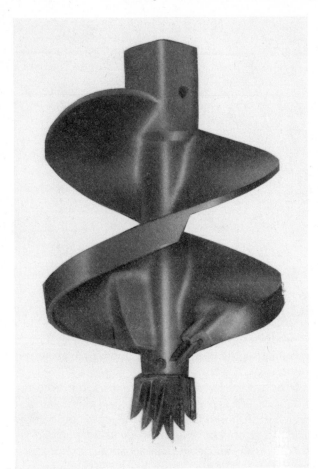

FIGURE 2.12.   Hughes MICON cast steel heavy-duty auger for hardpan and soft rock.

2.3.5.   Underreamers ("Belling Tools")

Underreaming tools (or "belling buckets") are available in a variety of designs. Generally, there are three main shapes of footing which can be cut with commercially available tools: the dome-shaped footings (approximately hemispherical, with a small cylindrical pilot depression at the bottom); the 60° underream (the sides making an angle of 60° with the horizontal; and the 45° underream. The choice of which to use seems to be principally dependent on local custom and experience, availability, and local building codes. Some contractors pefer the 45° tool claiming that because it removes a smaller volume of soil for the same bottom-bearing area it saves time in drilling and money in concreting. Others declare that the 60° tools, or the tool that cuts a dome, will drill faster and make a cleaner hole. (See also a discussion of these two shapes of underream in Subsection 3.3.3).

FIGURE 2.13.   WJF double-tooth step-type rock auger.

For large underreams, some contractors report that the bucket-type underreamer, which cuts a dome-shaped footing, carries out more soil per trip than the other designs and is therefore the fastest. This advantage has to be balanced against the cost of the additional concrete required by this shape of footing.

Some of the available underreamers cut a flat bottom, without any tendency to deepen the pier hole as they cut. Others cut a small sump at the bottom which is supposed to facilitate collection and removal of cuttings. Some underreaming tools have a tendency to deepen the hole or to pack cuttings under the bucket and create a false bottom of remolded and weakened material. This is something that must be guarded against by the inspector as the work is done. Most underreamers cut footings with a 6- or 12-in.-high (15 or 30 cm) cylindrical section at the bottom. Fig. 2.14 shows a 45° underreamer, and a 60° underreamer retracted and expanded.

Most underreaming tools are limited in size to a diameter three times the diameter of the shaft; and when larger bell diameters are required, either a larger shaft must be drilled or the machine-cut bell must be enlarged by hand labor.

2.3.6. Core Barrels

When rock becomes too hard to be removed with auger-type tools, it is often necessary to resort to the use of a core barrel. This tool is usually a simple cylindrical barrel, set with tungsten carbide (or other hard metal) teeth at the bottom edge. These teeth cut a small clearance both inside and outside the core barrel to facilitate flushing out cuttings, pulling the core barrel, and subsequently removing the core. The teeth are sometimes formed by building up the hard metal on a flame-cut, serrated edge of the core

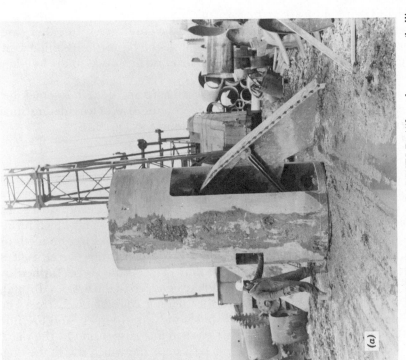

FIGURE 2.14. (*a*) CALWELD 45° underreamer (belling bucket). (*b*) CALWELD 60° underreamer (belling bucket).

FIGURE 2.15.   CALWELD rock core barrel with replaceable carbide-tipped teeth.

barrel, but more often are in the form of commercially available teeth which are set onto lugs formed on the bottom of the core barrel. In either case, the arrangement, spacing, and orientation of the bit teeth are important. The manufacturer usually furnishes detailed instructions as to the building up or setting of bit teeth.

For drilling holes where the rock is both hard and thick, it is sometimes necessary to use a core barrel equipped for the use of circulating fluid (either water or air). A tool of this sort is shown in Fig. 2.16.

2.3.7.   *"Calyx" or "Shot" Barrels*

For hard rock, which cannot be cut readily with the core barrel set with hard-metal teeth, a "calyx" or "shot" barrel can be used to cut a core of rock. In this barrel, the cylindrical skirt of the barrel extends upward some distance above the plate to which the drill stem is attached and which forms the top of the core barrel. The cutting edge of the barrel has no teeth or hard metal; it is composed only of the steel of which the barrel itself is made, with "feeder slots" cut into the skirt. The cutting is performed by chilled steel shot, which are poured into the hole, fed to the bottom through the slots, and are ground up under the rotating edge of the barrel. The grinding action of the broken steel shot grinds up the rock, and the fine steel dust and rock dust produced are washed into suspension by the water which is constantly pumped down the drill stem and which serves the dual purpose of keeping the grinding surface cool and also carrying off cuttings.

The suspended cuttings, rising through the narrow annular space around the core barrel (clearance being very small), emerge from this annular space into the water-filled hole above the core barrel and settle into the skirt at the top of the core barrel. This skirt is the "calyx" that gives this type of core barrel its name. Progress with this coring is slow, but a "calyx" barrel can be used to cut cores of the hardest rock. Operation of the "calyx" barrel is shown diagrammatically in Fig. 2.17.

2.3.8.   Multiroller-Type Bits; Blind-Shaft Bits

For cutting large-diameter holes through hard rock, particularly where the holes have to be very deep, adaptations have been made of the roller-type drilling bit which has been used for many years in the oil industry. The cutting elements in these bits are rollers furnished with teeth of hard metal; they drill by crushing hard rock, and both crushing and gouging softer rock, producing small cuttings which can be suspended in drilling fluid and flushed out of the hole. A sufficient number of rollers are disposed over the face of the bit to cover the entire area of the bottom of the hole as the bit rotates. The teeth on the rollers may be either pointed (chisel-shaped) or rounded, depending on manufacturers' recommendation or contractor's choice for the rock to be drilled. In general, the rounded teeth are recommended for drilling the hardest rock.

FIGURE 2.16.   Watson air-/water-circulating core barrel with carbide roller-bit cutters.

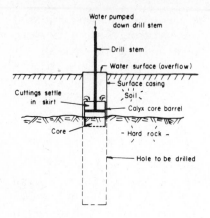

FIGURE 2.17. "Calyx" or "Shot" barrel, where cutting is done by chilled shot, under edge of rotating steel cylinder.

Multiroller bits are available for use with a "reverse-circulation"-type of rotary drilling rig, such as those used in the oil drilling and water well industries, or with an air-injection rig, in which the rock cuttings are carried up through the drill stem in a mixture of water (or mud) and air. Three types of circulation drilling—direct, reverse, and reverse with air lift—are shown diagrammatically in Fig. 3.9.

2.3.9. *Noncirculating Rolling Rock Cutters*

A roller-cutter-type rock bit that can be used in a dry hole, without water or air, has proved useful for locations where it is not practical to bring in a water supply or an air compressor. Efficient cutting with this tool requires high pressure on the rolling cutters, which is furnished by the crowd pressure of the drilling rig, plus the dead load that can be put in the weight basket on the top of the tool. To use this cutter, it is necessary to bring the bit out of the hole at intervals, and clean the cuttings out of the hole with a bucket tool or auger. This process is, of course, slower than making holes with a bit that uses circulating water or air; but for jobs with difficult access, and limited footage of rock drilling, it may offer substantial economy. The Atlantic noncirculating rolling rock cutter has been made in diameters from 14 inches (355 mm) to 84 inches (213 cm) (Fig. 2.19).

2.3.10. *Air-operated Down-Hole Bits*

One tool, the Atlantic "Cluster Drill," can be used with any crane-mounted drilling machine of suitable size. The "Cluster Drill" is equipped with multiple air hammers, impinging on the bottom of the hole and receiving their air supply through the drill stem. Each hammer has is own bit and the entire cluster rotates as it hammers. The cuttings are picked up by the high-velocity air stream at the cutting face, and are blown out of the work area and deposited in the calyx basket on top of the bit assembly. This tool will drill hard rock, even if it is broken up by clay seams, or with inflow of water. It is available in 24-, 30-, 36-, and 42-in. diameters (61, 76, 91, and 107 cm). A variation, the "Hammer Drill," uses a series of flat-faced hammers, reciprocating as the drill rotates. It is available in sizes up to 24-in. (61 cm) diameter.

The I-R Downhole Superdrill, made by Ingersoll-Rand, has a flat-faced rotary bit studded with hard-metal buttons. Initially developed for drilling through permafrost containing boulders, it drills hard rock readily. It is available in 20-, 24-, and 30-in diameters (51, 61, and 76 cm).

The "Cluster Drill" and the Downhole Superdrill are shown in Fig. 2.20*a* and *b*.

2.3.11. Raise Boring Bits

A raise bit is intended to drill a hole from the bottom up. There are relatively few applications for this technique in foundation work; but when opportunities for its use do occur, they may offer substantial economies, or even be the

FIGURE 2.18. Watson Tri-cone Roller bit with second-stage hole opener.

FIGURE 2.19.   Atlantic noncirculating rolling rock cutter.

only feasible way of putting a pier where it is needed. The applications that have come to the authors' attention have been in the construction of foundations for structures to be placed over existing cavities, as, for example, old mine workings, or natural caverns in karst topography. The cavity must, of course, be accessible for introduction of the raise bit.

In application of this technique, a small hole is drilled from the ground surface, to serve as a pilot hole for the raise drilling operation; the raise bit is then fitted to the bottom of the drill stem after the pilot hole enters the mine working/tunnel/cavity. Drilling then proceeds *upward*. A smaller amount of water and pump capacity are required than if the large hole were being drilled from the surface down and cuttings were being flushed out at the top. Manufacturers of raise boring bits are listed in Appendix A.

2.3.12.   Sidewall Grooving Tools

To facilitate development of the shearing strength of the soil or rock for a pier which depends upon sidewall shear for support, tools have been developed for cutting grooves or corrugations in the sidewalls of holes in materials which will stand without caving. These tools are usually shop-made, by the contractor. One form of grooving tool is shown in Fig. 2.21.

FIGURE 2.20. (*a*) Atlantic air-operated "Cluster Drill." (*b*) Ingersoll-Rand Downhole Superdrill.

FIGURE 2.21.   A shop-made sidewall grooving tool attached to the top of a drill bucket.

2.3.13.   Bailing and Cleanout Tools

Pier-drilling contractors are usually equipped with "bailing buckets," which are used to remove water from the hole when the drilling is complete. Some drilling buckets are so designed that they can be used also as bailers to remove most of the water in the hole quickly. Drilling contractors are often equipped with air lifts which can be lowered into the hole and used to pump out all but a very small amount of the water. In case it has not been possible to seal off water from entry into the hole, down-hole pumps, which can be left in place, operating until the concrete pour is started, are sometimes used.

2.3.14.   Mudding Tools

The mixing of prepared clay with water to make "drilling mud" is often a difficult and time-consuming task. If there is a great deal of it to be done, the contractor would be well advised to obtain mixing machinery such as that used in the oil fields. If it is only an occasional task, or if little mud is to be mixed at a time, the driller without a mud pit and a water jet can use the mudding auger shown in Fig. 2.22*a* for small-diameter holes. This auger has a tendency to become unstable when rotating at mixing speeds if its diameter is greater than about 3 ft (90 cm).

For larger-diameter borings, the "mudding bit" shown in Fig. 2.22*b*, which simultaneously cuts into the bottom of the hole and mixes drilling fluid and cuttings, has been found useful and is much more stable in rotation than the mudding auger.

2.4.   Casing and Liners

A distinction must be made between *casing* and *liners*. Casing is welded steel pipe of substantial wall thickness ("line pipe"). It may be either temporary

FIGURE 2.22. (*a*) Watson single-flight toothed mudding auger for simultaneously advancing hole and mixing mud slurry. (*b*) Overburden bit and mud-mixing cylinder for drilling in sandy overburden in mud-filled hole (reverse circulation). Bit diameter is 8 ft (2 m). (Photo courtesy of Girdler Foundation and Exploration Company, Clearwater, Florida.)

(left in place only to maintain a clean hole until concrete has been placed, then withdrawn) or permanent (becoming part of the finished pier). Casing must always be strong enough to resist the crushing forces that might be exerted on it by soil and water pressures before the concrete is placed. The importance of this consideration for large piers is supported by the fact that the resistance to lateral forces of a vertical cylindrical shell is inversely proportional to the *cube* of its diameter. There are several recorded instances of serious construction difficulties produced by buckling of large-diameter pier casings before the concrete was placed (2.4).

Commercially available sizes and wall thicknesses of "line pipe" are given in Appendix C. (Note that casing sizes are *always* specified by *outside* diameter.)

Because of its weight, steel casing has to be handled in moderate lengths. In some areas it is customary to use 10- to 20-ft (3- to 6-m) lengths of casing, attaching successive lengths by lugs and pins or bolts. In other parts of the country it has become customary to use telescoping casing, starting with an oversized hole and using successively smaller diameters of auger and of casing as additional depths of hole have to be cased.

Liners are intended only to act as forms to retain the concrete of the shaft. They are not expected to resist crushing forces or to contribute to the strength of the pier. Sixteen-gauge, galvanized, double-riveted corrugated metal pipe ("CMP") is often used. Because of the corrugations, the *inside* diameter of CMP should be the nominal diameter of the pier shaft (or larger).

Other fabrications, such as lighter-gauge welded pipe and even Sono-tubes, have been used under appropriate circumstances. Liners are protected from outside pressure by temporary casing or, where soil conditions are suitable, by being placed in a stable, noncaving, oversize shaft hole.

2.5. *Casing Vibrators/Extractors*

Electrically and hydraulically driven vibrators, initially developed for the rapid sinking of foundation piles and sheet piles in cohesionless deposits below the water table, have been adapted to the sinking and subsequent removal of the steel casings in the same types of soil. This type of machine uses a pair of counter-rotating eccentric weights, balancing out horizontal forces and imparting only vertical (or axial) vibrating forces to the casing to which it is clamped. The vibrating casing easily penetrates deposits of gravel, sand, silt, or mixtures of these soil gradations below the water table, and can be sealed into clay or soft rock below the cohesionless formations, permitting cleanout of the casing in the dry with an auger machine. Vibratory penetration of substantial thicknesses of clayey soil, of shale or harder rock, or of sands, gravels, and so on above the water table, may be limited or impossible.

Machines of this sort are available for clamping onto casings from 2 feet (61 cm) to 10 feet (3 m) in diameter. Where ground conditions are suitable, they can be a great timesaver. Figure 2.23 shows one machine of this kind driving a 2-ft-diameter (61 cm) casing in alluvial soil.

Vibrating driver/extractors are available both for external electric power source, and with self-contained diesel-fueled electric generator power supply. Several machines are listed in Appendix A.

2.6. *Kellys and Kelly Extensions*

Single-piece kellys are usually square, fluted, or splined driving shafts of solid steel, although hollow kellys are sometimes used to give maximum

FIGURE 2.23.   Foster Model 4000 Vibro Driver/Extractor on 80-ton Lima crane, driving casing through granular soil. (Photo courtesy of Harrison Drilling Company, Tulsa, Oklahoma.)

torsion capacity for minimum weight. Single-piece kellys are made to drill to depths of 70 ft (21.4 m) or more. When deeper holes are needed, some manufacturers use double or triple telescoping kellys, the inner square shafts sliding in larger hollow square sections. Other rig makers prefer to add pin-connected sections of drill shaft as needed. The Hughes LLDH machine is equipped with a telescoping kelly that will drill to a depth of 120 ft (37 m) without adding drill rods; the Calweld 70-ft (21.3 m) triple-telescoping kelly will drill to 190-ft (58 m) depth.

2.7.　Downward Force; "Crowd"

The combined weight of the auger bit or bucket and the kelly usually provides sufficient downward force for good, productive rates of pier-hole drilling through normal soil formations. But when the rate of penetration is slowed by harder formations, it becomes necessary to add downward thrust to maintain bit penetration at an economical rate. This thrust—usually called "crowd"—is generally added by the use of one or more hydraulic cylinders reacting against the weight of the machine carrying the drilling equipment. The amount of "crowd" truck-mounted rigs can apply is limited by the force that is sufficient to pick up the rear end of the truck and its load. The tools, kelly, and drill stem of a crane-mounted rig are usually very heavy, and, for many machines and in many applications, no additional "crowd" is required. Indeed, in most soil-drilling and many rock-drilling applications, the weight of bit, kelly, and drill stem (if any) may be more than is needed for good penetration; and often part of this weight has to be carried suspended from the crane and allowed to bear on the drilling surface only as needed. Some manufacturers augment this "tool" weight by use of hydraulic cylinders reacting against the weight of the drilling machine and the crane. Others take care of the requirement by adding weight to the tools, declaring that suspended weight will drill a straighter, more perfectly vertical hole than will any "crowd" mechanism.

In the drilling of large-diameter pier holes, a major reduction in the weight of tools resting on the drilling surface may be required in order to avoid overloading the individual cutting teeth of a bit or core barrel. When a very large multiroller bit is being used in hard rock, however, a major addition to tooth pressure may be necessary. The former has been accomplished by building a flotation tank into the top of the drilling tool (2.5), and the latter by weighting the drilling tool with steel punchings or with lead (2.6, 2.7).

References

2.1　Big Drills Twist 100-ft Caissons into Place, *CM&E*, May 1969, pp. 62–66. Seventy-nine casings up to 11.5 ft in diameter were sealed into rock or hardpan at depths of 80 to 112 feet for the 52-story IBM building in Chicago. Casings were left in place as required by current Chicago building code. Drill was Calweld CH on Vicon 3900 Manitowoc. Drill develops 400,000 lb-ft of torque. Casing barrels weigh 15–44 tons each. To limit pressure on cutting teeth, the casing is suspended from a second kelly. Two to four ft of penetration into rock is sufficient to seal off water. Manitowoc had to be reinforced, counterweight added, and lagging and sheaves changed to handle the cable. Contractor was Caisson Corporation, Chicago.

2.2   Caissons Twisted into Rock under Water Carry Elevated Highway over River, *CM&E*, April 1968 (6 pages—reprint from Hughes Tool Co., Houston). Three bridge piers for Interstate 435 over Missouri River near Kansas City, each with eleven 6-ft steel "caissons," the concrete and rebars extending 15 ft into rock. Eight-ft-diameter casing is twisted down by Hughes rig capable of exerting 1,000,000 lb-ft of torque; sand is then drilled out in a 6.25-ft hole down to rock, adding clay and water to make a slurry of the sand (patented "Case" method). Six-ft-diameter casing is then inserted, twisted 3 ft into rock, and emptied of slurry. Then 5.5-ft auger hole is drilled 15 ft into shale rock, and hole is cleaned out manually. Outer (temporary) casing is pulled after concrete has set. Contractor used Hughes CLLDH rig and tools from Case International, Chicago.

2.3   Tight Site for High Rise Foundations, *Construct. Digest*, April 1966 (2 pages—reprint from Calweld, Santa Fe Springs, CA). Hospital Service Corporation's 17-story building in Chicago's Loop district required 43 piers, 30-in. to 6-ft 9-in diameter, average depth 110 ft. Drill was Calweld 200 CH, rated at 360,000 lb-ft torque, mounted on Manitowoc 3900 Vicon crane. Caisson Corporation, Chicago, was foundation contractor.

2.4   Osterberg, Jorj O., Drilled Caissons, Design, Installation, Application, *Soil Mech. Lecture Ser., Found. Eng.*, Northwestern University, January–May 1968, pp. 151–208.

2.5   Downhole Float Controls Tools Coring Deep Caisson Shafts, *CM&E*, January 1966, pp. 67–71 (6 pages—reprint from Case International, Rochelle, IL). A detailed description of techniques, tools, and sequence of operations on the foundations of the 100-story John Hancock Building, Chicago. Deepest pier, 191 ft; most to about 150 ft; diameter 10 ft; for foundation pressures of 100 tsf on hard limestone.

2.6   Glidden, H. K., Floating Rigs Build Piers, *Roads Streets*, February 1965 (5 pages—reprint from Hughes Tool Co., Houston). Forty-four pier shafts for bridge across Sandusky Bay, Ohio. Holes 7 ft in diameter, average 60 ft below water and 10 ft into sound limestone. Bid cost $305 per cubic yard of concrete below water level. Special Williams (Hughes) rig, 88-in roller-bit cutting head, reverse circulation. Bit was weighted to 30 tons by adding hollow segments, filled with steel punchings, on top of cutter head.

2.7   Big Bit of Drilling Required for St. Louis Medical Center, *Construct. Digest*, November 7, 1968. One hundred and eight piers drilled with Williams (Hughes) CLLDH on Manitowoc 3900B. Holes were 36-, 42-, and 48-in diameter into hard dolomitic limestone. Bit was multiroller type, air cooled and cuttings blown out by air, and weighted to 55,000 lb by lead filling in skirt of bit. Contractor: Drilling Service Co. of St. Louis. Bells were made by hand labor and air tools.

3

Construction Techniques and Practices

3.1. Field Location and Plumbness of Holes

The first construction operation on a drilled pier job (after site grading) is the setting of stakes that tell the contractor where to drill holes. The drilled shaft contractor has to assume that these stakes are in the right places. But this does not guarantee that the centerlines of the pier holes will be in exactly the same places. Carelessness in starting the hole may cause the centerline to be moved a few inches, and stony ground, hardpan, surface rock, or improper tool design may cause the drill point to "walk off" location a few inches before it gets a good start. And the hole may be drilled out of plumb because the drilling machine mast or kelly (or crane attachment) has not been plumbed carefully, or, occasionally, because of ground variations.

The *actual* location and plumbness of a pier hole—as compared with its design location and plumbness tolerance—are the concern of several people:

1. The designer, who may have to specify remedial measures if the hole is too far off its designated location or too much out of plumb.
2. The surveyor, who must set the stakes in the first place and determine the as-built pier locations when they are finished.
3. The foundation inspector, who must check hole location and plumbness.

4. The contractor, who may have to do remedial work if either the pier hole or the finished pier will not pass inspection.

Most pier-drilling rigs are so built that it is easy to center the auger over a stake marking the pier location, and the "stinger"—a small pilot point or bit on the bottom of the auger—keeps the auger centered as the hole is started. However, when the hole must be started in hard, broken, or sloping rock (or even sloping hard ground), it may be difficult to keep the bit from "walking off' as the driller attempts to start the hole. In such cases, a pilot hole may be drilled first. In other instances, especially where the hole is to be drilled on a batter, a short cylindrical starting guide can be set to keep the bit or barrel on location. Starting guides are available from some drilling machine manufacturers, or they can be readily improvised in the field.

Vertical alignment (plumbness) is usually easily maintained, for the auger and kelly are very heavy and there is little tendency for a hole to "wander off' as it is drilled. Some machines depend on the weight of bit and kelly alone for "crowd" (downward force), governing the application of this force by regulating the tension in the wire line from which the kelly is suspended. Other machines have a hydraulic "crowd" mechanism to augment the weight of lighter tools and the kelly when drilling hard formations. Somewhat more care must be taken in the latter case to check plumbness of the hole, particularly in hard ground. A kelly of any significant length is quite limber, and forcing the kelly with a crowd does adversely affect the plumbness of a deep shaft. However, the authors do not believe that this consideration is often of structural importance, because the deviation from plumbness that occurs under usual conditions is usually of no significance as far as pier performance is concerned.

The greatest threat to plumbness is the presence of boulders, hard layers with a significant dip, or rock pinnacles. Where such obstructions produce serious deflections of drilling tools, corrective work, usually by jackhammer or the use of special tools and techniques, is required. Under these conditions, the completed hole can be considerably oversize. In difficult ground it may be definitely less costly to drill an oversize hole (which aids greatly in drilling through boulders) and use some extra concrete than to cope with hole location and plumbness difficulties.

Specification tolerances for centerline location and plumbness can make very important differences in construction costs and completion times. Specified tolerances should be as large as is compatible with realistic structural-design requirements and construction equipment and practices; and the specifications should be unambiguous and readily understood by contractor and inspector. Requirements of specifications and contract documents, governing hole location and plumbness, are discussed in Chapter 6, Specifications and Contract Documents. Designers and specification writers are referred to this chapter.

3.2. Depth Measurement

A final measurement of hole depth is necessary in every pier shaft; and in many, perhaps most shafts, several more depth measurements are required such as depth to rock; to bearing stratum; to obstructions; to water table; and so on. These measurements have to be accurate: they relate to pay items and to conformance with specifications. And they also have to be recorded. In the past such measurements were usually made by using a tape, and chalk (or other) marks on the kelly bar. Recent technology has made possible more convenient depth-measuring systems which are coming into wide use. One such unit (see Fig. 3.1) consists of two basic parts: (1) a measuring wheel, mounted in the upper mast section and rolling on the kelly cable; and (2) an electrical readout and control unit mounted close to the drill operator's controls. Depending on control switch settings, the instrument can read either depth to the auger point or penetration into an assigned stratum.

Operation of the unit is reported to be fast and simple, zeroing of the unit on starting the hole taking about five seconds, with continuous depth readings being displayed as drilling proceeds. The suppliers estimate that the use of the device will save 3 to 5 min for each time the driller would normally stop, clean, and measure, besides furnishing continuous data for logging stratum changes, and so on.

FIGURE 3.1. Depth Display Unit, Model DD-100. On the right is the measuring wheel unit (wheel rolls on the kelly cable) and on the left is the display unit (D.M.T., Inc., Denton, Texas).

3.3. *Readily Drilled and Noncaving Soils (Auger Drilling)*

3.3.1. *Straight-Shaft Drilling*

In soils in which drilled holes stand open readily, straight shafts are drilled very quickly, usually with an open-helix auger having two or three turns of a single flight. The auger may be equipped with either a knife-blade cutting edge or a series of replaceable cutting teeth, both of which can be faced with hard metal for longer life in abrasive formations (see Fig. 2.10). In general, the blade auger is used in uniform soils that are soft to stiff or very firm in consistency, while the toothed auger is used in decomposed or soft fractured rock, hard, cemented, stony soils, or very tough clayey soils that do not cut readily with the blade. Best progress is made with the blade when the auger is advanced fast enough to allow it to make a uniform cut but not to "corkscrew" itself into the soil like an earth anchor. A well-designed auger has a cutting blade on the side as well as the bottom to cut a clean hole and to make sure that the filled auger can be lifted without having to shear or tear the soil at its periphery.

Cutting blades and teeth should be kept sharp. Drilling with dull tools wastes time and money.

3.3.2. *Use of Water in Auger Drilling*

Some clayey or slightly clayey soils, particularly where the water table is deep, drill readily but have a blocky or crumbly structure that leads to sloughing or caving. It is common practice, in areas where these soils are encountered, to add water as the hole is drilled. This produces a smooth layer of moist, cohesive, remolded soil on the sidewall of the hole, often allowing the pier to be completed without casing and without sloughing. This practice is acceptable only for end-bearing piers.

3.3.3. *Machine Underreaming*

For most applications, any shape of underream—domed or sloped 45° or 60°, with a flat or a stepped bottom—is satisfactory if its shape is regular, its bottom is clean, and it stands without caving until concrete is placed.

The 60° underream (straight sides making an angle of 60° with the vertical) is often specified in building codes. This provision is probably due to the past perception that 45° underreams were subject to a flexural failure under high bearing pressures.

Many contractors feel that a 45° underream has a substantial economic advantage because the concrete volume required is substantially less and it can be cut much faster than the 60° bell. Further, for the same bell diameter, the height of the 45° tool is less than that of the 60° tool. For the larger-diameter piers, the 45° tool can therefore be used where a 60° tool would be too tall to clear the driving mechanism on the drilling rig. Because truck and crawler-mounted rigs are necessarily limited in height because of legal over-the-road height restrictions, tool height is a real limitation for drilling 60° bells with mobile rigs. A 60° bell specification can then require the use of crane attachments with greater rotary heights but much greater mobilization costs, where a mobile rig might be more economical for the specific job.

Some designers, ignoring the effect of confinement and of soil/rock modulus on the flexural resistance of a bell, continue to be concerned about overstressing a 45° underream despite its long successful use. Certainly at very high bearing stresses, provisions such as increasing the height of the toe (the outside cylindrical section at the bottom of the bell) may be used to maintain an adequate safety factor against concrete failure due to flexure under load.

An ongoing investigation at the University of Houston (3.1), of 7.5-ft-diameter 45° underreamed footings founded in very stiff clay, may provide a practical approach to an appropriate stress analysis. It is interesting that preliminary findings suggest that measured stresses are dependent on the properties of the soil supporting the bell as well as the geometry of the footing and the concrete strength.

In using the underreamer, the drill operator will establish a reference point on his kelly, or record the reading of his Depth Display Unit, as soon as the tool rests on the bottom of the hole. This will enable him to identify the true bottom on each trip into the hole. The underreamer is designed to complete its action in several "trips" but without deepening the hole or disturbing its bottom. In an operation where there is a tendency to pack cuttings beneath the underreamer, creating a false bottom of remolded soil, it may be necessary to clean out and reestablish the true bottom several times during the underreaming operation.

Some underreaming tools are made with a "reversed cutting" blade in the bottom of the bucket, which slides on the bottom of the hole during normal underreaming operations and picks up soil from the bottom when the reaming blades are retracted and rotation is reversed, thus permitting the true bottom level to be reestablished when necessary.

Underreaming cutters are activated by pressure from the kelly through a lever system housed in the underreaming bucket. If the drill operator tries to take too large a load into the bucket, progress will be impeded in several ways: (1) retraction of the cutting blades is made difficult; (2) spoils will be packed into the bucket so tightly that their removal is impeded and time is lost; and (3) complete plugging of the bucket may make it necessary to pull out against a partial vacuum in the hole. An attempt to lift out an overloaded bucket that has the reaming blades not completely withdrawn will result in raveling the sides of the shaft, or will hang up the bottom of the casing if casing is present.

Best production is usually obtained by coming out and emptying the bucket as soon as it is about one-third filled with loose—not packed—soil.

Some soils will stand open nicely in a straight hole, but cave and slough when an attempt is made to underream them. In this case it is sometimes economical to drill a straight shaft to the full underream diameter (or nearly full diameter, then cut a slight underream), pour a footing at the bottom, and set dowels and a corrugated metal liner of specified shaft diameter. The annular space outside the liner can be filled with sand, soil, or lean concrete, and the liner then be filled with concrete of the required strength. When this technique is used (assuming it is economically competitive to drill an enlarged diameter straight shaft), consideration must be given to the possibility of subsidence of loose backfill, especially if the pier is adjacent to a slab-on-ground floor, and to the alternative of a straight shaft with the same diameter as the bell—which is much less labor- and time-intensive.

3.3.4. Manual Construction of Underreams

Bells may be formed partially or completely by hand excavation in hard rock and where soils and weathered rocks tend to fracture, ravel, or cave during machine underreaming (2.7), but, in many circumstances, are best replaced by socketed or extended straight shafts. Pneumatic drills and spades are commonly used to excavate rock, hardpan, and stiff clay. Hand excavation also may be used in readily drillable soils when the bell is noncircular or is a larger diameter than can be cut by the largest available underreamers (3.2, 3.3). In rock and stiff clays, the excavation will usually stand open safely for this operation, but in some soils (e.g., in moist sand) shoring and bracing must be installed (or "spiles" driven) to maintain the roof as handwork proceeds. In the Los Angeles area, it is reported that hand-cut underreams have been common in some formations for diameters of more than 10 ft (3 m), and that the 1- × 4-in. wooden shoring is commonly left in place when the pier is concreted.

3.3.5. Multiple Underreams

Although rarely used in the United States, multiple underreams are sometimes used for the purpose of spreading the pier load over a wider zone in stiff soils or in layered soils where there are two or more strata of superior bearing capacity. Figure 3.2 shows a typical arrangement of two underreams, designed to mobilize the shearing resistance of soil over a larger area than would a single underream at the bottom. Additional advantages claimed (3.4) by its proponents for the multiple-underream pier (over a cylindrical pier offering the same sidewall shear area) are: (1) that because some of the shear is developed in undisturbed soil rather than in a softened "skin" adjacent to the concrete, and no reduction factor for softening or "smear" need be applied, the load-carrying capacity of the pier may be increased as much as 150%; and (2) that the volume of concrete for the multiple-underream pier is only 30–40% of that required for the same-diameter straight shaft.

Any contractor who can make piers with an underream at the bottom can drill the multiple-underream configuration—provided the soil is suitable

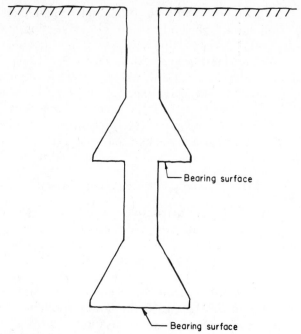

FIGURE 3.2.   Drilled shaft with multiple underreams.

for underreaming and stands without caving at the selected levels. Choice of underream levels and diameters is a design problem—the responsibility of the geotechnical engineer for the project—and may be open to serious question. The actual feasibility of this type of construction will have to be proven in the field for each project.

3.3.6.   Drilling Battered Holes

Battered pier holes are frequently required to provide horizontal reaction components for either compressive or tensile loads. Battered holes for non-foundation purposes—for drainage or for tieback anchors, for example—are not uncommon. Most truck-mounted drilling rigs designed for auger drilling can operate with the kelly at any angle between vertical and horizontal (and some even above the horizontal). Many crane-mounted machines can drill at batters up to 45°. Truck-mounted rigs designed for drilling with the drill bucket rather than the auger do not usually have the ability to drill with the mast tilted back. With these machines, batter is obtained by tilting the entire machine, either longitudinally or to the side, and, for practical purposes, the obtainable batter is limited to about 1 horizontal to 3 vertical.

The problems of drilling a battered pier hole are not confined to the setting of the angle of the rig mast and kelly. Setting the angle with the vertical is usually a simple matter, and can be done quite accurately. Setting the direction angle of the batter (i.e., the azimuth or directional bearing) is not so simple. The vertical angle will be indicated by a plumb bob and scale mounted on the rig mast. The direction, if it is to be set with any precision, must be determined by observation of the direction of the tilted mast by means of surveying instruments. Fortunately, a variation in the directional bearing of a battered hole is (in the authors' opinion) not usually of great importance, and a rough compass bearing—or even an "eyeball" setting—will generally suffice.

There may be a tendency for a battered hole to change its slope as it is drilled. If the soil formation is homogeneous (a rare condition!), there will be a tendency for the angle of the hole to become less steep as penetration proceeds, because of sagging of the unsupported middle of the kelly or drill stem. In layered formations, the deflection might be either upward or downward (or to either side), depending on the orientation and sequence of harder and softer strata. Fortunately, such deviations from straightness are usually of no structural significance for drilled piers or anchors. On the other hand, deviation of horizontal (or sloping) holes is extremely important when the hole is intended to tap an underground water source. This is true regardless of whether the objective is to intercept a natural aquifer or to tap a series of interconnected vertical wells designed to intercept groundwater flow. Alignment of such holes is a continuing problem for both the geotechnical engineer and the contractor.

It is important that design drawings and specifications should be reviewed for unnecessarily restrictive requirements for battered holes before a contract is signed. The drilled pier contractor should reconcile what he *can* do with what he *is required* to do before he commits himself. A designer who believes that his tolerances on batter, direction, and straightness are the maximum that can be allowed may be quite willing to permit larger tolerances in exchange for a larger-diameter hole or underream, or a deeper hole, or an increase in reinforcing.

In drilling a battered hole, there is more tendency toward caving than in vertical holes in the same formation; and, consequently, casing requirements may be more critical and underreams may not be as feasible as for vertical piers.

The concreting of a battered hole with an underream (or bell) at the end may introduce special problems. Because of an increased tendency to cave, the hole may have to be concreted more promptly than a vertical hole. The positioning of the reinforcing in the shaft is more difficult when the batter is very flat. Higher-slump concrete may be required, and the geometry of the underream may be such as to prevent complete filling with concrete, unless a special pipe is installed to vent trapped air in the upper part of the underream (see Fig. 3.3).

3.4. Hard Ground and Rock Drilling

3.4.1. Boulders, Cobbles, and "Hardpan"

For hard or stony soil and decomposed rock, auger bits and underreamers are usually set with hard-metal-faced ripping teeth rather than with cutting

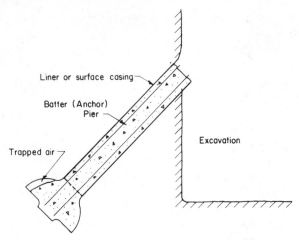

FIGURE 3.3.   Battered pier showing how air can get trapped during concreting.

blades (Figs. 2.11 and 2.13). These are very rugged and, used with a drilling machine having sufficient capability, will drill most cobbly soils, hardpans, and soft or decomposed rock with impressive speed. Hard boulders, however, will stop any auger. If a stone is not more than about one-third the diameter of the hole, it may be possible to pick it up with the drilling auger, or with a special short auger bit called a boulder extractor (Fig. 3.4), or sometimes with a "grab" (Fig. 3.5). If a stone is too large or too tightly held to be picked up thus, it must be broken up or loosened first. Breaking through obstructions can often be accomplished by dropping the kelly bar or the kelly equipped with a "gad" or "spudder" tool; or loosening (and sometimes breaking) can be accomplished with a "boulder rooter"—a wedging tool twisted by the kelly (Fig. 3.6). Other techniques for breaking up a boulder for removal are (1) to use a hydraulic rock splitter (Fig. 3.7), or (2) to inject expansive grout in small-diameter holes (see Appendix B) drilled in the boulder, or (3) by controlled blasting.

Loosened or broken pieces can then be picked up by an auger or a grab, or even extracted by hand. This is, however, a difficult and expensive procedure. A boulder-bearing formation is one of the most difficult to drill a good pier hole in, especially when the formation has a tendency to cave. Boulders impede the predrilling placement of casing by vibration, and make the use of slurry ineffective.

3.4.2.   *Layered Rock and Cemented Soil*

When the auger is stopped by relatively thin layers of hard rock (often interlayered with soil or soft or decomposed rock, or by soil layers that have become cemented), the techniques described above for boulders can be used to break up the layers so that they can be drilled with the flight auger (equipped with hard-metal teeth) (Figs. 2.11 and 2.12).

3.4.3. Drilling in Rock—Rock Sockets—Coring

Rock Augers. When pier holes have to be drilled into shale or weathered rock, penetration can usually be effected by using the toothed auger previously described or, in more difficult drilling conditions, by a special rock auger. Many designs of rock augers are available and the choice of which one to use depends on the type and condition of the rock, the ground conditions (layering, presence of water table, etc.), and the contractor's experience. Effective rock drilling requires skill and experience on the part of the driller. Too much pressure ("crowd") can break the auger teeth and too fast rotation can overheat and burn the teeth. With the correct speed and pressure, the teeth will fracture rather than grind or abrade the rock.

Two rock augers in current use are illustrated in Figs. 2.12 and 2.13.

Core Barrels. An auger must break loose or grind up 100% of the material within its diameter. A core barrel needs only to grind up the

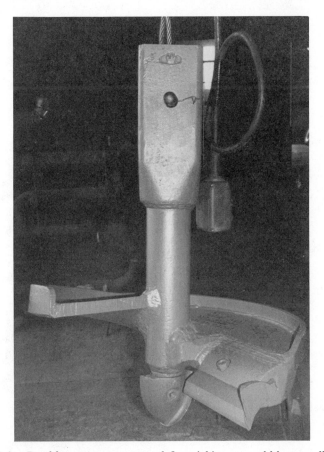

FIGURE 3.4. Boulder extractor—a tool for picking up cobbles, small boulders, broken ledge rock, and so on.

FIGURE 3.5. A hydraulically operated "grab," "rock bucket," or "cleanout bucket." This tool handles solids or fluids, and is used also for cleaning mud, laitance, and diluted or contaminated concrete from the top of a poured column, as in the illustration.

material in the annulus defined by the setting of its hard-metal teeth, usually limited to 1 or 2 in. of radius. Consequently, in rock that is so hard that a rock auger becomes uneconomical, hole is best made by using a core barrel and removing most of the spoil in the form of large cylindrical cores. Several varieties of core barrel are available with their choice depending on the same variables as mentioned for the rock augers and on the amount of hard-rock work to be done.

Simple Cylindrical Core Barrel. Except for use in very hard rocks, the core barrel usually consists of a simple cylindrical steel barrel set with hard-metal cutting teeth. The arrangement and spacing of the teeth are important: they must be set to cut sufficient clearance inside and outside the cylinder wall to accommodate the rock cuttings; they must be arranged to give complete coverage of the surface to be removed; and they must be in the right number to premit application of the necessary load per tooth through the drilling machine used. Moreover, their spacing should not be uniform, since this leads to chattering as the core barrel turns. A well-designed core barrel of this sort cuts out an annulus, leaving a pedestal of rock within the barrel.

When drilling with an auger machine and a core barrel of this kind, it is necessary to keep several feet of water in the hole at all times (although the water may have to be removed before the standing core can be removed). Speed of rotation should be governed carefully to comply with the core barrel manufacturer's recommendations (usually somewhere between 10 and 20 rpm). The pressure applied depends on the type of rock being cut, and the best pressure is usually found by experimentation. Too much pressure will usually result in excessive wear or tooth breakage. With the right machine, a good well-set bit, and a good operator, it should be possible to cut 30- or 36-in. (75–90 cm) core at a rate of 2 or 3 ft per hour even in hard rock.

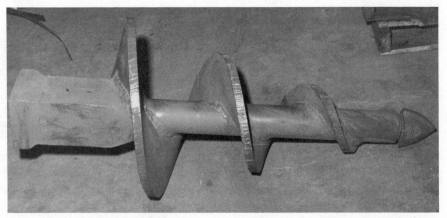

FIGURE 3.6. Watson Boulder Rooter for loosening or breaking up embedded boulders.

FIGURE 3.7. Darda Super-splitter (hydraulic) (Emaco, Inc.).

The coring of the rock by hard-metal teeth should be a cutting rather than an abrading process. To accomplish this the driller should allow the teeth to cool by lifting the bit every few revolutions. The core barrel is, of course, kept rotating during this brief cooling, and the raising and lowering of the core barrel serves the additional function of pumping cuttings out of the annular groove. Core barrel teeth must be kept sharp. The manufacturer will furnish recommendations for sharpening and building up worn teeth. Continuing to attempt to drill with a core barrel with worn teeth is inefficient, and has the additional disadvantage that whatever footage is made during this period is drilled without sufficient clearance and will give trouble when a core barrel with new or reset teeth is inserted.

The "Calyx" Barrel. Where rock cannot be cut by the hard-metal teeth of the core barrel described above, a "calyx" or "shot" barrel, with chilled (hardened) steel shot as the cutting medium, can be used successfully (Fig. 2.18). The "calyx" barrel has to be removed and emptied of its cuttings at appropriate intervals. This technique of drilling with chilled shot is slow, but can be used to core the hardest rocks. It will not work where rock is broken or fissured and the shot can be lost into cavities.

"Calyx" Barrel with Pilot Hole. A novel and very effective technique of accelerating the penetration of a "shot" barrel has been observed by the authors. When very hard rock is reached, a vertical hole, say 2 in. (5 cm) in diameter, is put down by jackhammer or by means of a small diamond core drill in the exact line of the annulus to be drilled by the core barrel. This hole is filled with chilled shot, and the drilling is started and proceeds in the usual manner. The presence of a fresh supply of chilled shot always at the cutting edge greatly expedites cutting action. (It might be mentioned that this technique works very well in drilling through concrete with heavy steel rebars.) This technique is illustrated in Fig. 3.8.

3.4.4. Removing Core from Hole

Usually a core will break off before the capacity of the barrel is reached. If it does not, the barrel is removed, a wedge attached to the kelly bar is driven on one side of the core to break it off, and the core barrel is lowered again over the loosened core. The core can usually be counted on to jam in the core barrel so that it can be lifted out of the hole, and can then be removed from the core barrel by tapping or hammering on the outside of the barrel while it is suspended.

If the core cannot be loosened and removed from the shaft readily in this manner, it can be broken up for removal in pieces by the use of the hydraulic rock splitter, the use of explosives, or the use of expansive grout, in one or more holes drilled into the core by jackhammer tools (Subsection 3.3.1).

3.4.5. Jackhammer ("Dental") Work

When a pier hole is centered over an irregular rock surface, so that one side of the bit bears on rock and the other on soil, or is over a cavity, or bears on a sloping rock surface, it becomes impossible to drill a straight

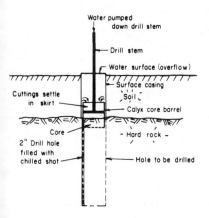

FIGURE 3.8. Shot (Calyx) barrel with pilot hole to feed chilled shot to cutting edge.

hole. Despite this difficulty, drilled piers with rock sockets may be the most feasible type of foundation, and in this case machine drilling has to be supplemented by putting a man down the hole with a jackhammer. In some cases it will be possible to take out just enough rock to establish a level surface that can be drilled by core barrel. More often, it proves necessary to complete the rock socket or underream by jackhammer work, supplemented by use of the hydraulic rock splitter or expansive grout; or, in rare cases, by blasting.

The use of explosives under these circumstances requires expert planning and performance to avoid risk of damage to other piers, pier holes, or bearing surfaces. Recent reports (1984) indicate that the cost of removing rock in this manner can be as high as $1500 per cubic yard, and occasionally even more.

The use of the rock splitter, or expansive grout, to break up the rock surface can be apparently very effective; but the cracked rock cannot be simply lifted out of the hole and still has to be loosened and further broken up before removal.

Another technique that may be used when a sloping rock surface is too hard to be leveled by the techniques described above is to clean the exposed sloping surface, pour a leveling course of concrete, and then drill down the leveled column. Whenever this sort of condition is found to be troublesome on a site, the geotechnical engineer and the structural engineer would do well to consider whether or not a pier set on, not into, the rock surface, but doweled against lateral movement by steel grouted into a smaller hole or holes in the rock, would serve.

This type of rock excavation in a pier hole is very expensive, with costs sometimes amounting to more than $1000 per cubic yard of material removed (1984). Rock conditions requiring such "dental work" are common where bedrock is of solution-prone limestone or dolomite.

3.4.6. Drilling with Circulating Fluid

Although a very large proportion of the pier holes drilled in the United States are drilled with auger-type machines and tools, supplemented in hard rock by the use of core barrels (Subsection 3.4.3) and sometimes manually operated pneumatic tools ("jackhammers"), for some regions and some pier designs an entirely different drilling technique is more favorable. This is the use of rolling-cutter bits, with circulating fluid to carry cuttings away from the work surface and out of the hole. The process is a modification of oil-well-drilling practice. The circulating fluid may be water, drilling mud (see Appendix B), air, or a water–air mixture. The circulation may be down through the hollow drill stem and up the annular space around the drill stem; or up through the drill stem, to return down the annular space ("reverse circulation").

With mud or water as circulating fluid, the cuttings are carried out of the hole and deposited, in one way or another, in a settling basin (commonly

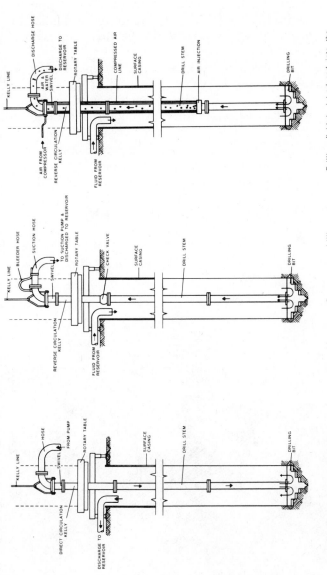

Drilling fluid is pumped down a hollow drill pipe, around the drill bit and back to the surface in the annular space around the drill pipe. Cuttings are carried to the surface by the flow.

The volume of drilling fluid required for a satisfactory rising velocity within the annular space depends on the volume of that space. This, in turn, determines the capacity of the pump needed.

In this counter-flow system, drilling fluid is introduced into the annular space around the drill pipe where it moves down, and around or through the drill bit. The fluid – along with excavated material – is then drawn back up to the surface through the inside of the drill pipe by a high capacity, low head, centrifugal pump. This system must be primed and the hole kept full of fluid at all times.

Drilling fluid is introduced into the bore hole as with the pumped reverse circulation method. The driving force to return the flow, and cuttings, to the surface is created by injecting air into the drill string. The composite mixture of fluid with air bubbles which is lighter in weight than the fluid outside the drill pipe forms a pressure differential thus creating the flow.

A proper fluid level must be maintained in the hole to provide the hydrostatic head pressure necessary.

FIGURE 3.9. Principles of circulation drilling. (Drawing courtesy of CALWELD, Inc.)

"mud pit"), from which the water or drilling mud can be returned to the circulating system.

Because reverse circulation produces a much higher rate of upward flow in the fluid carrying out cuttings, it has a distinct advantage in the drilling of large-diameter holes, an application where the upward fluid flow can be achieved by air injection, which simplifies the circulating system, and sometimes eliminates the need for mud slurry.

The advantages of the use of a circulating fluid system to carry out the cuttings are obvious:

1. It avoids the repeated trips to empty auger or bucket, and therefore saves time.
2. The drilling tools used with this system—rolling cutters, hammer drills, and so on—cut hard rock faster than the best auger bits; can often drill through boulders, rock pinnacles, or other conditions that would slow or even stop an auger; and thus avoid the necessity for dewatering and jackhammer work when difficult drilling is encountered.

When used with reverse circulation, this system is used to drill holes up to 20 ft in diameter in hard rock. A Hughes Combination Shaft Drill of this capacity is illustrated in Fig. 3.10.

FIGURE 3.10.   Hughes CSD-820 rig on bridge foundation, New York State.

Core Barrel with Circulation. For coring very hard rock, where there is excessive wear on the hard-metal teeth of the conventional core barrel, rolling cutters (adapted from oil-well-drilling tools) have been mounted on the cutting edge of the barrel, and cut sufficient clearance for the cuttings to be carried up the annular clearance space outside the barrel. With the work face kept clear of rock cuttings, the bit rollers are relieved of unnecessary wear and do not have to expend their energy on grinding up cuttings; and the speed of coring is much increased. In core barrels larger than about 24 to 30 inches in diameter, there is a settling chamber ("catch basket") on the top to catch the rock cuttings as they rise above the annular clearance space and settle out in the slower moving fluid (as described also for the calyx barel). An example of this design of core barrel is shown in Fig. 2.16. When smaller core barrels are used, the circulating fluid velocity is usually sufficient to carry the cuttings out, and the "catch basket" is omitted.

A very important advantage of circulation drilling using fluid, air, or both combined, is that the depth of a shaft can be far greater than the normal machine rating, since drill stem sections can be added without bringing the tool up out of the hole ("tripping").

3.4.7. Sidewall Grooving

Many tools have been developed for grooving or roughening the sidewalls of shafts or sockets drilled both into soil and into the rock. The use of these tools is often specified in some areas of the United States whenever the design used depends upon pier support by sidewall shear. Application of this technique seems to be a matter of local geology, local experience, and local availability. In the Denver area, for example, such grooving of pier holes drilled into rock is considered routine. In northern California, grooving of straight-wall piers drilled into hard soils is sometimes specified (3.5). Notwithstanding a growing body of load test data which demonstrate that rough-wall rock sockets support significantly higher loads than companion smooth-wall sockets (3.6), this technique is still considered in some areas to be technically unjustified and a contractor's nightmare, and consequently is not used at all. A shop-made sidewall grooving tool is shown in Fig. 2.22.

3.5. Drilling in Squeezing, Caving, or Sloughing Soils

3.5.1. Squeezing Ground

Occasionally, when a hole is being drilled in otherwise competent ground, a soft stratum, usually of clay or organic silt, will be encountered which tends to squeeze in and reduce the diameter of the hole while the drilling is being performed. To permit this action to continue while the hole is being deepened is to risk a reduction of hole diameter, with loss of ground and delays in drilling due to hangup of the auger bit. Whenever a squeezing

FIGURE 3.11. Watson Casing Twisting Bar (retracted position). This tool, attached to the kelly, is used for "screwing down" casing through squeezing or caving soil, for seating casing into rock or clay stratum, and for pulling casing.

condition is encountered, casing should be inserted and used to limit this action as far as possible. The casing may have to be driven or drilled ahead of the excavation to get it past the zone of squeezing soils. The twisting bar shown in Fig. 3.11 can be used for this purpose.

It should be remembered, however, that even though firm soil or rock is encountered below, and drilling may proceed below the cased-off zone, the squeezing soil will seal to the casing. This may result in a necked or soil-contaminated shaft due to the squeezing forces which are in excess of concrete pressure at the time the casing is withdrawn. A good rule would be to leave the casing in place in such formations.

Wherever squeezing ground is encountered in a pier hole (or any excavation), there is the potential for ground loss, causing subsidence close to the pier, and care should be taken that no damage to existing structures is caused. This sort of damage is a potential source of litigation.

3.5.2. Caving or Sloughing Soils

Cohesionless soils (gravels, sands, silts) below the water table will usually cave when drilled by an auger. Exceptions to this occur when such soils are cemented. Conversely, unsaturated cohesionless soils which are fine enough to mobilize significant negative pore pressure (suction) are often stable during drilling, at least for limited periods. However, these intergranular stabilizing stresses decrease with an increase in the degree of saturation and are lost upon complete saturation. Unsaturated coarse-grained soils such as gravel tend to cave unless intermixed with, or have particles coated with, clay or silt-clay-size particles. Unsaturated silts, fine sands, and other sand-silt gradations with moderately low permeabilities can be caused to cave when saturation is induced by collapse of soil structure caused by vibration during the drilling operations.

3.5.3. Stabilization by Drawdown

The drawdown procedure often works well in fine and well-graded sands. As water level is drawn down *outside* the hole, seepage forces around the periphery of the hole are reversed; capillary forces are set up which tend to keep the sidewalls from caving; and, if drilling is performed smoothly and without delay, a clean open hole can often be established all the way through the caving stratum. Suitability of soil conditions for this technique can be established from the test borings, correlated with observations of the actual pier drilling as it proceeds. Where soil conditions are favorable, it may be possible to greatly expedite a pier-drilling job by the use of groundwater drawdown, as compared with the best progress that could be obtained if casing were used to seal off the caving strata. However, where highly compressible subsoils are present, use of this technique can produce substantial ground subsidence, and the possible effects on nearby structures should not be overlooked.

3.5.4. Casing the Hole

The most usual and direct means of coping with caving cohesionless soils is to case off as the hole proceeds through the caving formation. This can be a tricky operation. It is usual to drill the hole larger in diameter than the outside diameter of the casing to be used. This clearance may be as little as 2 in. for a 24-in. casing, or as much as 6 in. for a 48-in. casing (depending somewhat on the kind of soil being drilled).

Even in caving soils, it is often possible to drill the hole quickly and insert the casing freely before caving has started or has reached troublesome proportions; but conditions will be encountered where this cannot be done. When pore-water pressure is sufficiently high, sands and silts cannot be removed to extend the hole past the bottom of the casing without causing the soil to "run in," and serious loss of ground may result. Under these circumstances it may prove impractical to struggle with alternate drilling and driving of casing (for driving or twisting will be necessary whenever the caving ground seizes the casing). Several techniques are available, however, one or another of which may serve to keep the hole open while drilling is being performed and allow the casing to be either advanced with the hole or dropped in freely when the shaft hole is completed. The first of these is to keep a positive head of water in the hole during drilling—that is, keep the hole full of water, or nearly enough full that hydrostatic pressure is sufficient to prevent run-in of sand at the bottom—and to alternate drilling or screwing the casing down ahead of the boring, with drilling (usually with the bucket auger) to remove the soil within the casing (but not ahead of the casing). A second method is the use of a mud slurry in the hole to balance the hydrostatic pressure of groundwater outside. A third is to draw down the groundwater level around the hole, usually by the use of deep wells suitably arranged and pumped down to a level well below the bottom of the proposed drilling, thus setting up capillary forces in the dewatered soil.

When casing is practicable but installation is difficult because of caving or seizing formations, the use of a vibrating pile driver is sometimes very effective, particularly in granular soils (Fig. 2.23). In clayey soils, the casing will sometimes build up a cake of clay both outside and inside the casing, so that penetration is greatly slowed after a few insertions and removals.

3.5.5. Sealing Casing into Rock

When a pier is to be founded on rock or drilled into rock, there is often difficulty in making a satisfactory water cutoff at the rock contact. Where the rock is soft or weathered, a good seal can usually be effected by driving the casing, or by twisting or rotating it while pushing it down. If the plain end of the casing will not penetrate far enough to make the seal, flame-cut teeth on the bottom may be effective for cutting into soft or weathered rock, and the cuttings formed will help in establishing the water seal. For harder rock, it may be necessary to face the cutting teeth with hard metal (3.7, 2.1).

Under some conditions sealing of the bottom of the casing may be extremely difficult or even not possible with the equipment available. This occurs when the rock is hard and the surface is sloping or irregular, or when there is a layer of cobbles or boulders immediately over the rock surface, or when the bedrock is fissured or cavernous. Under such circumstances, the casing may be lifted; the water inflow may be stopped by balancing it with a sufficient head of water or mud in the hole; and a mat of a few inches of lean grout may be placed on the bottom by tremie or pumping. After this has set up, then casing can be drilled through it into the rock.

In instances where a good bottom seal cannot be made as described above, grouting with cement or chemicals might be considered, with the objective of forming an impervious bottom zone into which the casing could be sealed. One large pier-drilling contractor reports successful sealing under such circumstances by using a proprietary material called "Sealite," mixed with cement and packed into leaking cracks, leaving a small pipe in a crack for pressure relief. The relief pipe can be grouted later.

If none of the above techniques effect a good seal, then it becomes necessary to complete the hole in the wet (filled with mud slurry or water), then concrete it (carefully) by tremie or by pumped concrete (Subsection 3.12.4).

3.5.6. Stabilization by Mud Slurry

The use of a mud slurry to keep the hole open can be applied to a wide variety of soil conditions. This technique has been adapted in part from oil-well-drilling practice; however, whenever there is sufficient clay in the soil being drilled to allow it to be stirred, with the addition of water, into a fairly stable slurry, the mud slurry used in foundation drilling, unlike oil-well-drilling mud, may not be composed principally of commercial bentonite, but rather of the soil of the site. For less clayey soils, commercial bentonite (Aquagel, etc.) can be added. For examples of mud use, see Refs. 3.2, 3.3, 3.8, 2.2.

To mix most clayey soils with water sufficiently to make a good slurry, the drilling contractor will need some special tools to do it, as well as special tools for drilling in a mud-filled hole (Fig. 2.22a and b).

To drill a self-mudded hole, it is usual to drill a starter hole, the removed soil being replaced with water. The drilling is then continued through the water, mixing the soil drilled with the water, until a slurry of suitable consistency is produced, with more water added when the mixture becomes too viscous. If the soil being drilled has little or no clay, it may be necessary to add commercial bentonite to keep the soil in suspension. Carefully designed and controlled commercial bentonite should *always* be used in conjunction with tremied or pumped-in-place concrete placed in water.

To a person not accustomed to its use, the efficacy of a viscous or heavy drilling fluid to keep a hole from caving is surprising. It will allow drilling a hole in sand—dry, moist, or saturated—with straight, clean walls. It can even enable the driller to underream in a clean sand below the water table, without any caving or sloughing of the roof of the underream. This performance is not automatic. It requires skill and experience on the part of the driller—and understanding of the technique and confidence in the contractor on the part of an engineer who authorizes or approves its use. Whatever drilling is done in a mud-filled hole is done blindly; the results can be judged only by behavior of the tools, augmented by probing. The sidewalls and underream cannot be seen, even with the use of an underwater television camera. It would be possible, of course, to have the surfaces inspected by touch, by a diver, but the authors know of no instance of this being done. A final, indirect check on the hole can be made by comparing its design volume with the volume of concrete that is placed in it. This is a very important part of the inspection process *and it must be done carefully*.

When starting a hole that is to be mudded (as well as holes under many other circumstances), it is advantageous to begin by drilling an oversized hole and setting a short (8–15 ft) (2½–4½ m) piece of oversize surface casing. This helps the starter hole to retain water and slurry, and prevents caving and fall-in at the surface caused by operations of men and machines.

In instances where the bottom of the hole (or the underream, if any) is to be in a noncaving material, the hole should be drilled oversize down to the lower stratum. Casing somewhat larger than the pier shaft can then be set through the slurry, and seated into the lower stratum (see Subsection 3.5.5). In this operation, the slurry acts as a lubricant, facilitating insertion and sealing of the casing. When the casing is firmly seated and sealed, the slurry can be bailed or pumped out. The remainder of the hole can then be drilled in the dry (which is much faster than drilling in slurry).

3.5.7. Control and Cleaning of Slurry

When a hole is drilled by bucket or auger, using a slurry as a stabilizing agent, each time a load of spoil is lifted out of the hole the volume of spoil removed must be replaced by more slurry (or water in the case of a self-mudded hole).

In a slurry-stabilized hole, the slurry must perform several important functions—and refrain from performing some unwanted ones:

1. Its density must be sufficient to exert enough hydrostatic pressure on the side surfaces of the hole to prevent sloughing, caving, or collapsing.

2. But its density must never become so high that the slurry in the bottom of the hole (where it is densest because of the settling out of cuttings) will mix with, or intrude into, or refuse to be displaced by concrete being placed by tremie or by pumping. *Caution*: this phenomenon, if not prevented, can result in seriously weakened concrete at the bottom of a shaft, and has been identified (3.9) as the cause of sand inclusions and leakage in retaining walls built by slurry-trench methods.

3. The slurry's viscosity must be *high* enough to prevent most suspended soil particles (sand, silt) from settling out during drilling operations.

4. But its viscosity must be *low* enough that it will be readily displaced by concrete placed by tremie or by pumping.

5. It should gel ("set up") when agitation stops, so that suspended sand and silt do not settle out during interruptions in drilling operations, or after completion of the hole, while awaiting concreting. And its gel structure must break down, and fluidity be restored, when agitation is resumed.

To measure and control these properties of a clay slurry, three rather simple instruments are usually required: a Marsh funnel for viscosity, a sand content meter (with a 74-micron screen), and a "mud balance" for bulk density measurement. In addition, in some soil and groundwater situations, pH-measuring equipment is required, since soil minerals or groundwater solutes may react with the drilling clay to alter its viscosity.

There are not (in 1985) any standards in the United States for either viscosity or sand-content limits for drilling slurry for pier hole drilling. Probably most drill foremen believe that they can tell from machine behavior and slurry appearance, "feel", and behavior when sand content, density, or viscosity is outside good operating limits. However, in some circumstances—large or deep augered pier holes through sand, for example—the sand content and viscosity of the slurry near the bottom of the hole can be vastly different from that at the surface, where it can be observed; and the difference is unfavorable toward the successful placement of concrete. When the bottom density, or sand content, or viscosity is too high, there is serious danger of trapping a compressible pad of clayey sand on the bottom, or of contamination of the tremied or pumped concrete by inclusions of loose sand.

For large or important jobs in sandy soils, testing of slurry properties during drilling, and cleaning the slurry of sand before concreting, or before re-use, are often required; and these will sometimes become specification requirements. See Subsection 6.2.3-5 for slurry specifications.

Another consideration is that the material removed in making the hole ("spoil") must be wasted—hauled away and disposed of as specified in the contract documents and in conformity with local antipollution regulations. A clay-stabilized slurry is liquid or semiliquid, and a suitable disposal location for it cannot be found as easily as one for sand. Consequently, slurry so loaded with sand that it cannot be re-used is an operating expense, not only for the cost of replacement of the drilling clay, but also for the extra disposal cost. When holes are being drilled in sand, a clay slurry usually becomes too loaded with sand for further use after two or three holes have been drilled with it, and must be wasted or else be separated from the sand. A commercial machine for cleaning slurry of sand is illustrated in Fig. 3.12.

A special form of drilling fluid, called "Revert," has been extensively used in the water-well-drilling industry. This is a material which, when mixed with water, makes a viscous fluid which will suspend and carry out cuttings as effectively as a mud slurry. After a predetermined time, however, the fluid reverts to about the viscosity of water, and is therefore very easily displaced by concrete (see Appendix B for source).

3.5.8.   *Concrete-Rebar and Concrete-Sidewall Bond in Slurry-filled Holes*

Experiments have shown that concrete properly placed in a slurry-filled hole, where the slurry viscosity and sand content have been controlled

FIGURE 3.12.   Caviem Desander/Desilter for construction site slurry. (Photo courtesy of Pileco, Inc., Houston, Texas.)

properly, makes a good bond with reinforcing steel, and with the sides of holes in soil, despite the thin coating of slurry that wets their surfaces (3.10, 3.11). This is not the case for smooth-wall rock sockets, where reductions due to slurry drilling may be as much as 25 percent (3.6). Where the drilling of piers designed for sidewall shear is common practice, it is not unusual to permit the use of drilling mud provided the required pier capacity can be demonstrated by full-scale load tests on a designated number of completed piers.

However, in the soft (and permeable) rocks of South Florida, tests have shown that the substantial buildup of drilling mud cake, attributable to the permeability of the rock formations, results in a major decrease in the sidewall shear support developed by the pier. It would be prudent to avoid the use of drilling mud in drilling holes for piers where sidewall shear is to be relied upon for support, in any soft and permeable rock formations (3.12, 3.6).

3.6. Piers Through Water into Soil or Rock

Drilled piers are frequently used for waterfront structures, bridge piers, and other construction in both shallow and deep water. When soil conditions below water are suitable for the use of this type of foundation, casing can be drilled or driven down to where it is firmly held by the soil, and then drilling can proceed just as for an installation on land below a high water table (2.2, 3.13). Where the casing can be sealed into a suitably impervious formation, it can be dewatered and the drilling and underreaming can proceed in the dry; or, if dewatering proves impractical, the mud-drilling technique (Subsection 3.5.5) can be used to complete the pier hole. A discussion of concrete placement in such cases is given in Subsections 3.12.5 and 3.12.6.

3.7. Use of Expansive Cement

The use of expansive cement in the concrete of drilled shafts has been proposed, with the purpose of increasing sidewall friction and thereby increasing the bearing capacity and reducing settlement under load. One study investigating this technique has been reported (3.14) and research is continuing. Preliminary results (1983) indicate that the use of a suitable expansive cement mix in the shafts of piers through very stiff clay can increase the skin friction of the shafts by 25 to 50 percent, with a reduction of settlement under load of as much as 50%. One corollary effect of using the expansive mix is a reduction of the strength of the concrete in the upper part of the shaft, where expansion is not inhibited by development of restraining soil pressure as the concrete expands. At the time of writing

(1985) this technique must be considered experimental. However, it is probable that further study will lead to its advantageous use under some circumstances, and that guidelines and specifications for the technique will be developed and published.

3.8.   *Grouted Piles*

A bored pile (drilled pier) with a grouted base and sidewall friction area, which, when drilled in predominantly sandy soils, is claimed to have bearing capacity greatly increased over that of an ordinary pier with the same dimensions, has been developed by a West German firm. This pier has been used in Europe, North Africa, and the Middle East.

This pier has a reinforcing cage extending to its full depth, with pipes attached to enable grouting the base and the sidewall surface. Attached flat to the base of the cage is a circular flat steel diaphragm, similar to a "flatjack," with its inner space connected to one set of grouting pipes. The other set of grouting pipes has grout outlets distributed along the part of the cage where the pier is supposed to develop support from sidewall shear (Fig. 3.13).

The cement grouting procedure is as follows:

1. After the cage has been placed and centered, the pier is concreted in the usual manner (by tremie or pumped concrete).
2. Two days after concreting, grout is pumped into the sidewall grout pipes, under high pressure; it ruptures the green concrete, fills any voids between pier and soil, densifies loosened soil, and increases the lateral stresses against the pier surface.
3. Two days after the grouting described above, grout under high pressure is injected into the flatjack at the bottom of the pier, and the pressure is increased until the peripheral seal of the flatjack is broken and the grout flows out and compresses and penetrates the surrounding soil.

Details of this system can be obtained from the developers (3.54).

3.9.   *Underpinning*

Drilled pier-type foundations are particularly adaptable to underpinning problems for several reasons. Pier holes can be drilled immediately adjacent to existing foundations, with relatively little vibration and without ground displacement. Pier construction is rapid, so that underpinning construction can be completed in a minimum of time. Under most circumstances, the underpinning pier can be taken down to a safe bearing level with comparative

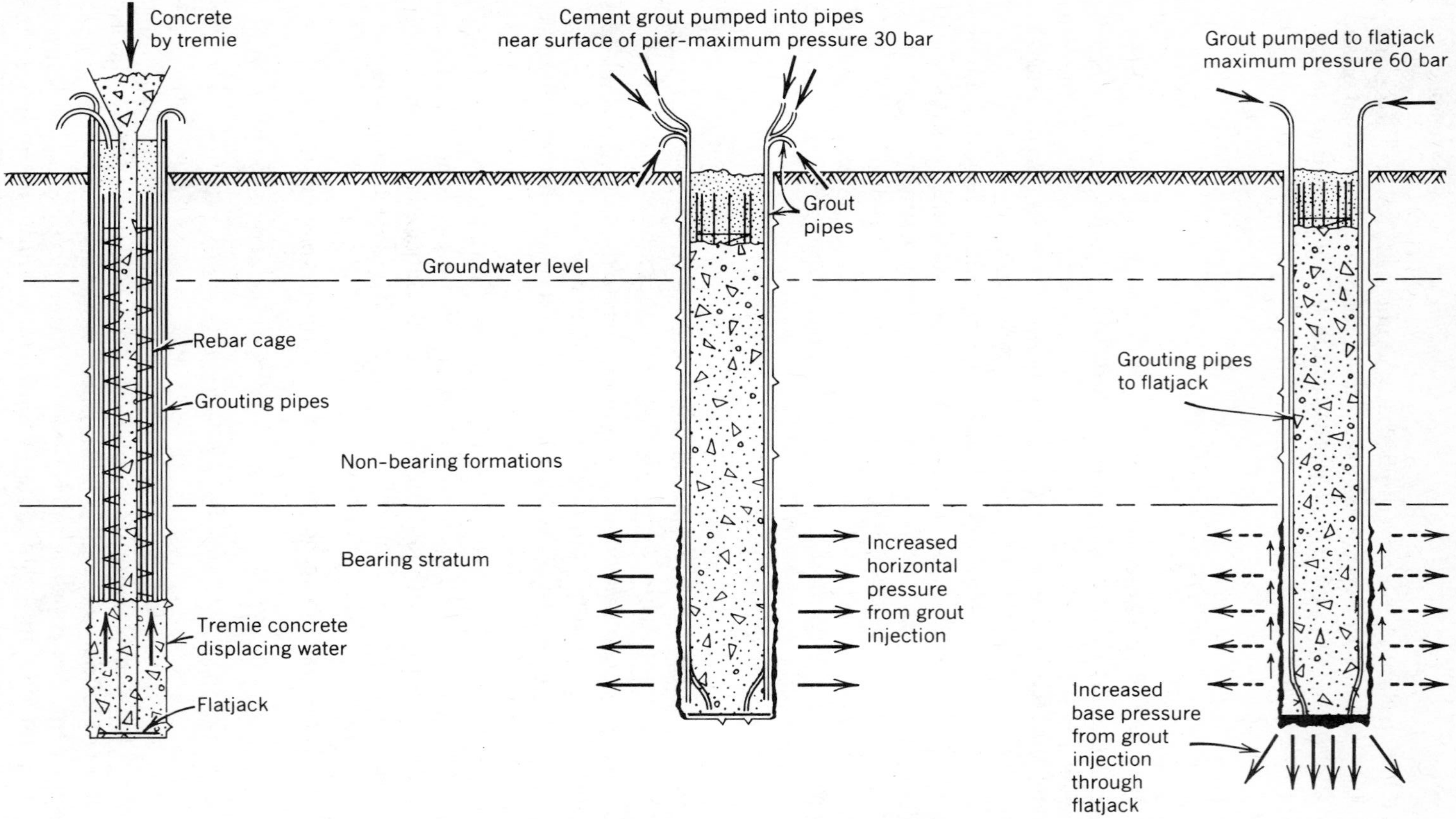

Concrete by tremie

Cement grout pumped into pipes near surface of pier–maximum pressure 30 bar

Grout pumped to flatjack maximum pressure 60 bar

Grout pipes

Grouting pipes to flatjack

Increased base pressure from grout injection through flatjack

Increased horizontal pressure from grout injection

Groundwater level

Non-bearing formations

Bearing stratum

Rebar cage

Grouting pipes

Tremie concrete displacing water

Flatjack

FIGURE 3.13. Bauer System Grouted Pier.

ease. In some instances, underpinning piers can be combined with drilled pier diaphragm walls, forming part of the basement wall for the adjacent new structure. For interior underpinning work, as in remodeling or in restoration of support to settling interior columns, specially mounted pier-drilling machines for operation under low-headroom conditions are available in many cities.

Precautions must be taken against loss of ground during drilling operations for underreaming. When squeezing or caving ground is encountered, careful use of casing may be required. If casing placed in predrilled holes is left in place, however, consideration must be given to prevention of subsidence, which may be caused by soils moving into voids left in the annular zone surrounding the casing. Under some circumstances, the use of drilling mud may be adequate to prevent caving or hole collapse, even when the soils drilled are cohesionless. When the underpinning shaft is drilled through stiff clay, the hole will usually stand open long enough to permit completion and concreting before ground movement can develop.

3.10. Preparation of Hole for Concreting

3.10.1. Cleanout Techniques

Any pier hole which depends upon bottom-bearing for its support must be cleaned out carefully before concrete is poured. It is possible with some bucket auger bits, and in some soil or rock, to drill a clean bottom. However, in some underreamed holes and in some straight shafts, cuttings or "slop" on the bottom of the hole may require handwork to produce a good bearing surface. This necessitates descent of a workman into the hole, with the appropriate cleanout tools, and with the mandatory safety provisions; and a subsequent descent by the inspector to see that the job has been done properly. Although this is common practice in high-capacity rock-base piers, this is an expensive and (sometimes) dangerous procedure, and alternate methods should be used whenever possible. On any job where sloppy bottoms are a problem, airlift equipment will usually do a satisfactory cleanup job, and may be sufficient to avoid the risk of descent into the hole.

3.10.2. Shape of Bottom of Hole

Different tools cut bottoms of different shapes. Some bottoms are nearly flat, some are dished, and some are rounded or conical in shape. As long as the hole is cleaned out properly the profile of the bottom makes no appreciable difference to the bearing capacity of the pier. In the authors' opinion, there is no objection at all to the use of tools which cut a "stepped" bottom, or which make a pilot hole a few inches deeper than the nominal bottom of the pier.

3.10.3.   How Clean Is Clean Enough?

Cleanness is a question which is usually left up to the judgment of the geotechnical engineer, or the foundation inspector acting under the instructions of the geotechnical engineer. It is the authors' opinion that the inspector alone, without special instructions which will depend on the materials penetrated in the drilling and the soil or rock that comprises the bearing surface, may not be qualified to exercise judgment in this matter. If there is difficulty in getting a clean bottom in a pier hole, the matter should be called to the attention of the geotechnical engineer at once; he should make an inspection and he should give definite instructions as to what should be done. A small volume of dry or plastic cuttings on the bottom of the pier hole will make no difference in bearing capacity. If, however, there is as much as an inch or two of soft "mud" in a plastic condition, which will not be displaced to the outer circumference as concrete is poured, but which is compressible enough to permit measurable settlement, then this would be unacceptable, and special means would have to be devised for getting the hole cleaner. It must be remembered that, where the bottom of the hole is in very impervious soil such as clay, the cuttings or mud trapped under the concrete as it is placed may not have an opportunity to compress appreciably in the brief time that it takes for the concrete to set up, and therefore they could constitute a compressible layer which might produce measurable postconstruction settlement.

Cleanup of sidewall smear for piers drilled in soil is, in the authors' experience, only rarely needed or required. However, cleanup may be required for smooth-wall rock sockets drilled in shales and other rocks whose cuttings may produce smears similar to bentonite (see Subsection 3.5.8). Sidewall cleanup is readily achieved by hosing or jetting with water.

It is sometimes the practice to "stabilize" the bottom of a hole containing a few inches of water, mud, or "slop" by adding enough cement or gravel–cement mix, to absorb the water and form a paste or soil–cement mix so rich in cement that it will set up sufficiently to be practically incompressible under load. To form such a mix, the added dry materials will have to be thoroughly blended with the material in the hole. This may be attempted by inserting the auger in the hole and mixing by a few turns of the auger, without allowing the blade to drill any deeper. The authors have seen this method used, and do not know of any postconstruction settlement that can be attributed to it; but they regard the practice as questionable. If this in-place bottom stabilization is used, it must be done carefully; and it should be avoided whenever possible. The use of a smaller sump filled with "stabilized" material is preferable.

In instances where bottom softening is due to seepage of small amounts of water from above, a "mud slab" (a few inches of concrete placed immediately on completion of the boring) will protect the bottom soil and facilitate later cleanout just before the main mass of concrete is placed. But when water is rising under pressure from below, a mud slab could be lifted or broken by water pressure and could be responsible for later sudden settlement.

For piers designed for sidewall shear support only, bottom cleanout is of minor importance. In such cases, the only cleanout is performed with the drilling tools, and manual cleanout and visual inspection of the bottom are often omitted entirely. Of course, care must be taken that enough sidewall contact for the concrete is available above any disturbed material remaining on the bottom, and that whatever "slop" is present is not fluid enough to intrude or dilute the concrete above the level at which shear support is counted on.

3.10.4. Bad Air or Fog in the Hole

Bad air or poisonous fumes can come from many sources, and for safety reasons this possibility must be taken into account whenever any person enters the pier hole (see Subsection 4.5.1). A common occurrence, particularly during winter construction, is the formation of fog in the hole, so that inspection of the bottom is impeded or prevented (3.15, 3.16, 2.4). When either bad air or fog is present, it should be dispersed by pumping in air before and during descent into the hole by any person. Any hole that has been left open for an extended time should be assumed to have bad air and be flushed out before being entered.

3.11. Dewatering—Groundwater Control

To assure good bearing and high-quality concrete, not more than 2–4 in. of water (depending on base diameter) should be left in a pier hole when concrete is placed. The technique of sealing off water with casing has been described in Subsection 3.5.5. In instances where water has entered the hole before the casing seal was effected, the hole can be bailed almost dry by the use of a bailing bucket attached to the kelly. In other instances, water in the hole can be taken care of by direct pumping, or pumping from a sump hole drilled in the bottom of the pier hole. If the rate of water entry is slow, concrete placement can then follow in the normal manner.

Where the rate of infiltration is too large to prevent excessive water from accumulating in the hole before concreting, it is sometimes feasible to use a pump fitted with a length of small-diameter pipe ("stinger") on the intake. The intake pipe is kept on the bottom and pumping is continued until a plug of concrete about equivalent to the stinger length is placed. The pump is then lifted out and the pour is completed. For situations where groundwater infiltration cannot be controlled by pumping, tremie placement, after the water has been allowed to rise to its static level in the hole, is a viable solution. As introduced in Subsection 3.5.3, groundwater control can also be provided by depressing groundwater levels by installing and pumping from deep wells or well points, or from adjacent open pier holes. This technique is not recommended if there is the potential for subsidence and, under some circumstances, may not be effective or economically feasible.

Excess water left on the bottom of a pier hole can produce a surprising volume of clean gravel instead of concrete at the bottom of the pier. Larger amounts can be quite disastrous when the concrete is placed by free fall. There is one report of 2 ft (60 cm) of water in the hole resulting in 8 ft (2½ m) of sand and gravel at the bottom of the pier (3.17).

3.12. Concreting

3.12.1. Prompt Placement of Concrete

Prompt placement of concrete in the prepared hole, as soon as it is completed, is a necessity if troubles from caving, squeezing, water entry, and difficulty in pulling casing are to be avoided.

The manner in which the concrete is placed is also of the greatest importance and improper placement techniques are the greatest cause of defective piers.

3.12.2. Free Fall

Experience and experiment have shown that concrete falling freely does not segregate provided the falling concrete does not strike the sides of the hole or a reinforcing cage (3.17, 3.18). A drilling contractor advises that concrete was successfully placed for the pier foundations of the Federal Building, Cleveland, Ohio, using a free fall of as much as 167 ft (51 m).

It is not sufficient (except in shallow holes) to discharge the concrete directly into the hole from the chute on a mixer; this will almost invariably result in the concrete's striking the side of the hole (or the cage) as it falls. And the authors do not consider satisfactory the practice of having a workman hold a shovel so as to deflect the concrete stream from the chute so that (maybe) it falls straight down instead of striking the cage or the side of the hole.

It is necessary to have a hopper centered over the hole, with a bottom spout designed to concentrate the falling concrete in a stream of small diameter compared to the hole diameter, or else to place the concrete through a tremie pipe or "elephant's trunk" (a canvas conduit) discharging at or near the bottom of the hole. In order to completely eliminate segregation, emptying of the hopper should be avoided until the hole is filled with concrete. This is impracticable when dealing with large piers requiring several loads of ready-mix, or when casing must be withdrawn in sections; but the interval between discharges should be kept as brief as possible to avoid the development of a layer of laitance on top of the concrete in the hole before the next truckload is placed (refer to Subsection 3.12.10).

3.12.3.   *Concrete Design Considerations*

Slump and maximum aggregate size must be carefully controlled in order to be sure that either free-falling or tremied concrete will flow freely between reinforcing bars and completely fill the space outside the cage. Sometimes it is advantageous to govern aggregate shape as well as size. Some authorities specify gravel rather than crushed stone for coarse aggregate in concrete to be used in reinforced piers, to avoid the "harshness" associated with crushed stone mixes (3.18).

The slump (a measure of fluidity) of a concrete mix is critical to the integrity of a drilled pier. Where the slump is too low or the mix too harsh, there may be arching of the fresh concrete in the hole or around reinforcing steel, leading to voids or honeycombing in the concrete and/or incomplete bonding of the rebars. Placement of such concrete during withdrawal of casing may also lead to seizing in the casing and separation or necking of the concrete column during withdrawal of casing, followed by intrusion of soil or mud.

Prior to the early 1970s, following the practice of formed concrete placement, slumps of about 4 in. were commonly used in many areas of the United States and full depth vibration was often specified to prevent arching and incomplete bonding of rebars. Subsequently, higher slump concrete has become common and current practice calls for as much as 6- to 7-in. slumps for conventional concrete placement. With the increased fluidity of such concrete, vibration is not required although vibration in the upper 5–10 ft of the shaft (where the consolidation stress is least) is sometimes used to assure proper bonding of the rebars and dowels.

If concrete is to be placed in water or slurry-filled holes, using either tremie or pumping in-place techniques, special concrete mixes are required. These employ smaller aggregate sizes and higher slumps than conventionally placed concrete. For example, maximum aggregate sizes of ½- to ⅝-in. and slumps of 7 to 8 inches are characteristics of tremied concrete mixes. By comparison, pumped concrete may employ an even smaller aggregate size and 8–9-in. slumps.

In the authors' opinion, the concrete placed in a drilled pier should be designed to provide some reserve other than that strictly required for the support of the design loads. This approach is explicit in the design procedures prescribed by ACI 318 where the target concrete strength (f'_c) represents a 9% exclusion limit approach. It is also their opinion that it is prudent not to specify an f'_c less than 3500–4000 psi; there is typically little cost savings with lower strengths, and greater than normal strength variations have sometimes been observed with low-strength mixes.

These items should be covered in the plans and specifications (see Chapter 6).

High slump can—and must—be combined with suitable strength. This can be accomplished by adding cement—at the expense of shrinkage—or preferably by the use of suitable additives. It *must not* be done by adding water to a well-designed mix with low slump.

Particular caution must be used in the acceptance of ready-mix deliveries in very hot weather. The concrete will take its initial set sooner, and sometimes a delivery man will add water to be sure he can discharge the load satisfactorily—with disastrous results to concrete strength. On the other hand, to go ahead and place stiffening concrete with a lower slump than the design calls for can easily result in voids and honeycombing. Hot-weather concreting calls for special care in design, delivery, and placement. It may be necessary to use a retardant in the mix. The inspector should have the authority to require the use of a retardant wherever needed.

3.12.4.   Tremied and Pumped Concrete in Dry Holes

In many piers where reinforcing cages are used, the cage diameters will be small enough to obstruct the free fall of concrete. Where inside diameter of cages is less than 2 ft (60 cm), tremie placement can sometimes be used, but pumped-in concrete may be required.

Tremied or pumped-in concrete will not be subject to segregation in a dry hole, provided the bottom of the pipe is always kept below the surface of the concrete, and the tremie pipe is large enough to avoid excessive friction.

3.12.5.   Placement Below Water

Before tremied or pumped-in concrete is used, care should be taken that there is no water entering the hole; otherwise, the hole (or casing) should be filled with water (or in some cases drilling fluid) to the very top. With the water level at the top, any leaks occurring will move water out of the pier, not into it. No dilution or segregation of concrete can then occur due to water entering while the hole is being filled with concrete. This is an important precaution.

When concrete is placed below water by tremie, a common source of dilution or segregation is water in the tremie pipe before concrete placement is begun. The pipe should be empty (except for air) when the pour is started, and it should be resting on the bottom of the hole. This means that a temporary closure of some kind will have to be fitted to the open end of the pipe before it is inserted in the pier hole. When the concrete in the pipe has reached a level where the concrete pressure exceeds the water pressure at the bottom, the closure must open or come off readily as the filled pipe is lifted slightly.

A special closure of plywood has proven satisfactory, and is light enough to float out on top of the rising concrete—if it does not get hung up in the reinforcing cage. If the plywood disk does not float out, it is sufficiently incompressible that it would not be expected to weaken either the footing or the shaft appreciably. Another closure that some contractors have found satisfactory is an inflated basketball. The expedient of blocking the end of the tremie pipe with rags, or with anything that would be likely to produce a serious weakness wherever it lodged, should never be condoned.

Some tremie pipes are made with openings in the side, shaped so that falling concrete will not escape through them, but concrete will flow out of them freely when concrete rising outside the pipe has reached their level. This speeds the pour.

3.12.6.　The Tremie Pipe

The authors' experience indicates that a tremie pipe should be at least 8 in. (20 cm) in diameter, and that 10 or 12 in. (25 or 30 cm) is better. Some authorities recommend that the diameter of the tremie pipe be at least eight times the size of the maximum aggregate.

Several instances of serious weakening of concrete by contamination from contact with aluminum pipe have been reported (3.19). The use of this metal to convey fresh concrete should be avoided under all circumstances.

3.12.7.　Advantages of Pumped Concrete

Although pumped concrete is usually more expensive than tremied, it has certain potential advantages that may prove more important than the cost differential. When it is available at all, it can usually be obtained in strengths up to 6000 psi (420 kgf/cm^2). Where the need for it can be anticipated, the design value of f'_c can be increased. Because discharge can be at the bottom of the hole until the entire column has been concreted, continuity of the concrete without segregation can be assured. In some instances this will justify the use of a smaller concreted shaft size, and the saving in concrete volume will go far toward compensating for the increased cost per cubic yard for the pumped concrete.

3.12.8.　Vibration During Placement

As described in Subsection 3.12.3, vibration of concrete in drilled piers is sometimes specified, presumably on the assumption that this provision will assure that the concrete makes complete contact with reinforcing bars, as well as avoiding voids in the part of the pier where hydrostatic pressure of the wet concrete is least. Down-hole vibrators suitable for use in deep piers are readily available. However, it has also been reported that vibration of concrete inside a temporary casing will sometimes cause packing and wedging of sand and gravel outside the casing, resulting in considerable difficulty in pulling the casing. Similarly, vibration of a harsh or dry, low-slump mix, may cause the concrete to seize the inside of the casing so that it cannot be pulled without lifting and separating the concrete.

There have also been reports of segregation produced by use of vibrators, and some authorities limit its use to the top 4–5 ft of concrete in a pier, or to the depth of the bottom of the reinforcing cage (if any). It is the authors' opinion that the use of vibrators in drilled pier concrete should be limited to the upper 5 ft of the pier, and that flow of concrete through rebar cages, and so on, should be governed by control of slump and aggregate size and shape.

3.12.9.   Pulling the Casing

When a hole has been cased to seal out water (as distinguished from casing placed to protect personnel or to prevent minor sloughing of sidewall from contaminating the bottom), the casing must not be disturbed until enough concrete has been placed to produce a higher concrete pressure at the level of the casing than the water (or slurry) pressure outside the casing at that level. If this precaution is neglected, water will enter the concrete column when the casing seal is broken (Fig. 3.14b). The result will be a weakened column, and in some cases an open void or a zone of aggregate washed clean of cement.

It should be emphasized, however, that the lateral pressure of fresh concrete is not necessarily the hydrostatic pressure computed for a liquid weighing 145 pcf. The higher the slump, the more nearly the lateral pressure will approach the theoretical hydrostatic pressure. Recent experiments (3.20) showed that concrete with a 9½-in. slump (obtained by using a superplasticizer) exerted full theoretical hydrostatic pressure for the full height of the formwork (29 ft), while a 5½-in. slump mix, with measured lateral pressure integrated over the entire depth, indicated only a little over half of that total.

If there are cavities of substantial volume outside the casing, concrete will flow out from the casing to fill them, and the level of the concrete in the casing may drop far enough for it to no longer exceed the outside water pressure. It is critical, then, that attention must be given to maintenance of concrete level in the casing as it is pulled, particularly when cavities may have developed during the drilling. (In some cases, such cavities can be "packed off" with clay soil during drilling and before casing is placed.)

A circumstance which is especially likely to give trouble is a sealed-off zone with artesian pressure above the initial bottom seal of the casing. If an artesian zone exists and is vented to ground surface by clearance between casing and hole, there can still be a positive head of concrete against the entry of water into the column as the casing is raised. A sealed-off artesian zone can be vented to the surface by drilling a small hole to its level, alongside the casing (Fig. 3.15d). Care must be taken, if this is done, not to disturb the casing's bottom seal.

When telescoping casing is used, with the lowest section sealing off groundwater, it is necessary to place enough concrete in the lowest casing section to be sure that, when the casing is pulled and the bottom seal is broken, the surface of the concrete level does not fall below the bottom of the next larger casing (Fig. 3.15). This will depend on the relative positions of the top of the bottom casing, the bottom of the next section, the level of groundwater outside the bottom section, and the volume of water to be

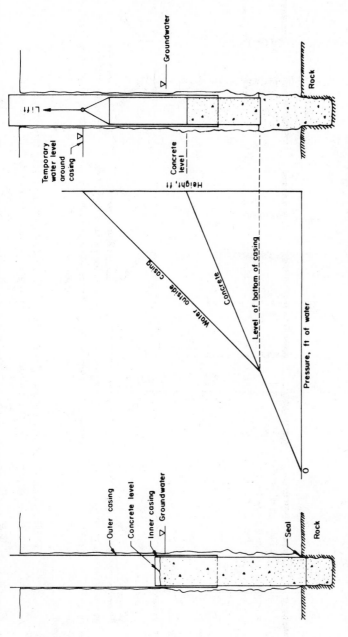

FIGURE 3.14. Telescoping casing showing relation between groundwater pressure and concrete pressure.

81

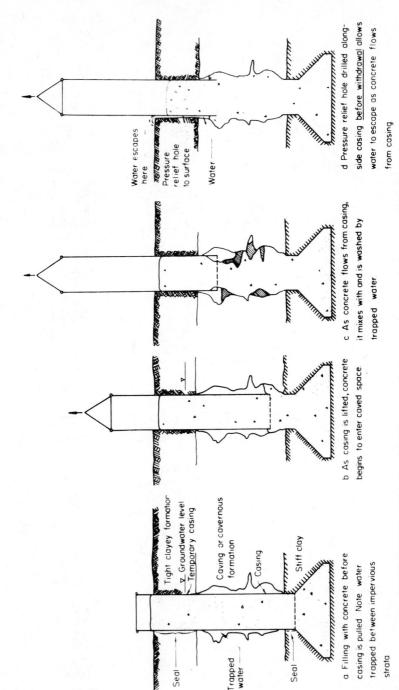

FIGURE 3.15. This sketch illustrates (*a*, *b*, *c*) contamination of pier concrete by groundwater, and (*d*) avoidance of contamination by maintaining concrete pressure during pulling of casing.

a Filling with concrete before casing is pulled Note water trapped between impervious strata

Seal
Trapped water
Seal

Tight clayey formation
▽ Groundwater level
Temporary casing
Caving or cavernous formation
Casing
Stiff clay

b As casing is lifted, concrete begins to enter caved space

c As concrete flows from casing, it mixes with and is washed by trapped water

d Pressure relief hole drilled alongside casing before withdrawal allows water to escape as concrete flows from casing

Water escapes here
Pressure relief hole to surface
Water

displaced outside the lower casing and below the bottom of the next section. If there has been much caving during the drilling, or if the lower formation is cavernous, the concrete level should be watched carefully as the bottom section is lifted; and more concrete should be added if it appears that concrete level is about to fall below the bottom of the next section, thus admitting groundwater. As the casing is lifted, the inspector should require monitoring of the distance down to the concrete surface, and compare it with the expected (computed) drop for the distance lifted. The need for this precaution is illustrated in Fig. 3.15.

Another circumstance that can lead to serious mistakes in pulling casing is the presence of a cased-off stratum of soft "squeezing" soil, which can produce "necking down" of the pier shaft when the casing is removed.

When a reinforcing cage has been placed inside the seated casing and the casing has been filled with concrete and is then pulled, forces can be developed in the cage that it was not designed to withstand, especially if the concrete is stiff or harsh. In order to fill the annular space exposed below the rising casing bottom, as well as any additional cavities in the sidewall, the concrete inside the cage must slump and flow downward, putting the cage in hoop tension; and the concrete between casing and cage likewise flows downward, adding to the vertical compression force on the cage.

When the pier hole has been drilled through caving or cavernous ground, the slump of concrete as the casing is pulled can be very substantial. It is not uncommon for a spirally wound reinforcing cage to fail at this stage by wracking, the twisting action expanding the spiral turns while shortening the longitudinal members. A positioned reinforcing cage, which is expected to remain stable while the casing comes out and the concrete goes in, can shrink vertically and disappear from sight entirely, leaving the top of the shaft unreinforced and the observers mystified (Fig. 3.16).

This phenomenon, as indicated above, may become serious when the space to be filled outside the casing is substantial. It can also occur as an underream is being filled with concrete through a cage resting on the bottom. It is amplified by the use of too harsh a mix, by the use of concrete with too large aggregate, by too close bar spacing, or by carelessly made or inadequately designed ties in the cage. It can be minimized by use of concrete with a slump of 6 or 7 in., and eliminated by supporting the cage from the top while concrete is being poured. In cages which are long enough to rest on the bottom of the hole, distortion (wracking) can be prevented by welding horizontal hoops around the cage at appropriate intervals.

Casing should be pulled in an axial direction, never by an eccentric pull. If the drilling rig is being used to handle the casing, a twisting bar can be used to break the seal; if a crane, a short jerk may be used; otherwise the pull should be smooth, without jerks. Concrete level should be maintained so that it never falls below the bottom of the inner casing until the level of

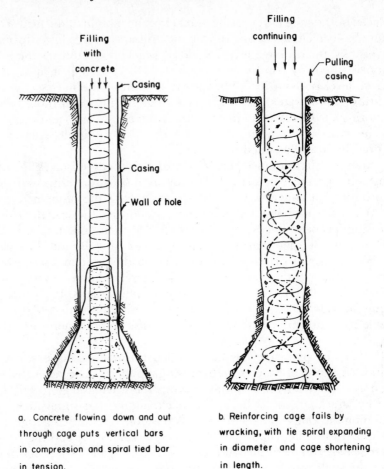

a. Concrete flowing down and out
through cage puts vertical bars
in compression and spiral tied bar
in tension.

b. Reinforcing cage fails by
wracking, with tie spiral expanding
in diameter and cage shortening
in length.

FIGURE 3.16. Distortion of an inadequately tied rebar cage by weight of low-slump or harsh concrete.

the surface casing (or design level for top of pier, whichever is lower) has been reached.

If the concrete has a suitable slump, and the casing is pulled promptly as described above, no trouble should develop during the pulling. If the concrete is too harsh, or if the casing is jerked or pulled too fast, the concrete column can be pulled apart below the bottom of the casing, with disastrous results to the pier. When this occurs, the reinforcing cage may be found to have risen several feet by the time the casing is out of the ground.

Many contractors have found that a vibratory hammer/extractor is very effective, not only for driving casing but extracting the casing smoothly without lifting the concrete. However, it must be cautioned that this technique can cause densification of cohesionless soils, thus interfering with smooth casing removal.

Special attention should be given to the ground adjacent to the casing during pulling. The development of subsidence at that time indicates a strong probability of intrusion of soil or water into the newly exposed column of concrete.

3.12.10.   Interrupted Pour

Although the placement of concrete in a pier should be continuous from start to finish, this is not always possible. When a partially filled pier is allowed to stand until concrete has taken its initial set, the exposed surface requires attention before more concrete is placed, to assure good bond between the old and the new concrete. Laitance should be removed, and the surface should be roughened and slushed with a 1:1 cement grout just before the pour is resumed. Where the reinforcing cage prevents direct access to the surface, it is imperative that all laitance be removed.

3.12.11.   Uncased Pier Through Water

An ingenious technique has been used to produce smooth-finish concrete piers through water without the necessity of leaving costly steel casing in place and subject to corrosion (3.13). The procedure is to drill the pier as usual, setting a smooth steel casing (not a corrugated or galvanized liner) to contain the concrete, but using a second casing, at least a foot larger in diameter than the other, centered with the pier and set far enough into the sea bottom to hold firmly, and securely braced in place. The annular space between the two casings is filled with clean sand, and concrete is placed in the inner casing. The inner casing is immediately pulled with a vibrating puller, using care to get a smooth vertical pull. The annular mass of sand retains the heavier concrete without any tendency to slough or intrude in the pier. After the concrete has set, the outer casing is pulled (loosening the sand annulus by jetting if necessary); the sand drops off; and the pier is exposed, with a smooth surface.

3.13.   Defective Piers

3.13.1.   Kinds and Causes of Defects

Defects in completed piers are probably no more common than defects in any other type of deep foundation. However, in most pier-supported struc-tures, a column load is supported by a single pier, whereas a group of piles would be required to support a similar load. For this reason, a single defective pier is more likely to result in structural damage or failure than is a single defective pile or a proportionate percentage of piles scattered through the project. This is one of the reasons why, in Chapter 4 of this book, the necessity for thorough and competent construction review and inspection has been emphasized. If these functions are adequately performed, any construction defects should be avoided or detected and remedied before a pier is completed, and the completed pier should be adequate to carry the design load without suffering detrimental settlement or deflection. However, experience shows that completed piers do occasionally turn out to be defective; and there have been several instances of near-disasters, sometimes avoided more by good fortune than by care and vigilance (3.16, 2.4).

The types of defect that can occur and the conditions that can lead to these occurrences have been reviewed by Baker and Kahn (3.16); they may be summarized as follows:

1. Inadequate concrete strength, caused by segregation during placement, by improper placement in cold joints, or by delivery of poor concrete to the site.

2. Displacement or contamination of the concrete due to improper casing withdrawal techniques or by caving or collapse of walls of hole or underream.

3. Voids or discontinuities in the shaft or underream, caused by use of concrete which is too stiff; by collapse of casing or liner; by squeezing in of soft formation; or by hangup of concrete in the casing while it is being pulled.

4. Inadequate bearing, resulting from inadequate soil investigation and evaluation, from error in identification of the proper bearing material while the hole is being drilled, or even, occasionally, from an unauthorized change in dimensions of the bearing area or omission of the underream.

Of all of the above, the improper concrete placement techniques and the use of concrete of too low slump (nos. 2 and 3 above) are clearly the most common causes of pier failure.

If investigation and design were always perfect; supervision always competent and adequate; inspection always continuous, experienced, and alert; and the contractor's personnel always expert and conscientious—if all these conditions prevailed, all of the time on a project—then there would be no defective piers.

3.13.2. *Verification of Pier Integrity—Checking Completed Piers*

For important or settlement-sensitive structures, and when large loads are to be carried on drilled pier foundations, the geotechnical investigation, construction supervision, and inspection should be performed meticulously. Also, however, consideration should be given to the advisability of postconstruction investigation to confirm the continuity and integrity of the completed piers on such projects. Not every such project will need this kind of check. When the engineer can be assured that piers have been drilled through formations that give no trouble, and that they have been adequately inspected and filled with concrete of the correct slump from a reliable source, with no water in the holes and with the work done by experienced and conscientious contractor's personnel, he can be reasonably certain that the piers are sound and continuous. But if any of these conditions deviates from the ideal, and particularly if there has been trouble in getting a dry hole or if the concrete has been placed under water (tremied or pumped in), then a check on the condition of the completed work becomes advisable.

Occasionally circumstances arise where it is necessary to verify the integrity of a drilled pier foundation system or of one or more piers suspected to be potentially deficient. Where a verification program is prescribed before start of construction, special concrete quality monitoring techniques can be employed. Usually, one or more small-diameter pipes are cast in the concrete shaft and are used to deploy nuclear, sonic, or other geophysical devices to provide continuous measurements of a specific concrete property (3.23).

The bulk density of the concrete is usually derived from backscatter measurements through cast-in-place sleeves and currently is the most common geophysical verification technique employed. Down-hole nuclear density instruments are manufactured by Troxler Electronic Laboratories (Model 1351). This approach was recently successfully used in the United States to verify the integrity of piers drilling in 30 ft of water to support a bridge structure (3.20) and is reported to be widely used in Caracas, Venezuela for verifying piers drilled to support high-rise buildings.

Down-hole propagation of seismic p- and s-wave velocities has also been induced in a shaft to provide an assessment of concrete quality. Direct transmission measurements can be made by casting velocity transducers (receivers) in the concrete. Alternatively, a surface-mounted receiver can be used to measure reflected waves (3.21). A variation of this technique has been pioneered commercially by the Birdwell Geophysical Company and involves a probe containing a sonic pulse generator and a transducer which detect and signal the arrival of the generated wave. The travel time and travel distance are electronically processed to provide a velocity profile. By casting two pipes in the shaft and lowering concurrently a sonic source and a detector, cross-hole velocities can also be obtained in large-diameter shafts and have the advantage of "seeing" more of the concrete. Note that concrete velocities are related to elastic moduli, and moduli to compressive strength.

Other types of in-situ geophysical measurements have also been infrequently used to detect concrete anomalies. These include resistivity and gamma-ray measurements (3.22). These geophysical systems, however, are not usually economically viable.

Most often, verification requirements stem from observed anomalies, construction difficulties, incomplete inspection, and specification noncompliances. Under these circumstances, the appropriate verification method is likely to be determined by what part of the construction needs to be verified. For example, if the concrete strength is suspected to be less than specified, concrete cores would be required for compression testing. If the shaft integrity is questionable, then a variety of verification methods in lieu of or in combination with coring are available. These include the in-situ tests previously described (using percussion drilling to make the access holes), and nondestructive testing.

At this time, the most direct postconstruction check on pier dimensions and soundness is obtained by diamond coring and a core-hole position survey.

The larger the diameter of the core, the more information can be obtained and the more expensive the test. For this application, the authors do not recommend the use of cores smaller than NX (2⅜ in. or 60 mm); larger cores, up to 6 in. (15 cm) in diameter, have been used. The choice of core size, of the number of core holes per pier, and of the individual piers to be investigated, depends on the reasons for making the test in the first place. The construction events or difficulties (if any) which led to a judgment that test core borings should be made should be evaluated to determine what defects may have resulted that should be looked for. For example, a 10-ft-diameter (3 m) pier for a very important building, in a situation where there was a strong suspicion of the presence of defective concrete, was tested by 11 diamond core borings (3.23). A pier for moderate loads, installed under circumstances that had not led to any suspicion of its integrity but where general policy dictated testing of a selected percentage of the finished piers, might be tested by only one diamond core boring.

In any case, where diamond core borings are used to test pier concrete, the work should be done by the most skilled contractor available. His equipment should be in first-class condition, and his drillers must be experts in diamond coring. In the authors' opinion, this requires a negotiated contract; and a price per hour, with a provision for payment for diamonds used, is fairer to the contractor than a price per foot of core. Diamond coring in concrete is difficult work at best. When zones of poor concrete, or of washed gravel, are encountered, difficulties and uncertainties are multiplied. An unskilled driller (or a worn bit) can produce results that make conditions look much worse than they really are. The problem is too important and the results too critical to justify trying to save money on the test-boring contract.

Diamond coring can be supplemented by other test procedures that often can expand considerably the information yielded by the cores and drilling records. The borehole camera (3.24, 3.25), usable in NX and larger holes, can be used to determine the extent of cavities or of layers of uncemented material—information often not determined accurately from cores. Automatic caliper logging, seismic or sonic velocity logging, and gamma-ray logging have been used to advantage. Inclinometer logs have been used to determine straightness and plumbness of test borings (conditions which are rarely determinable by direct observation) (3.23, 3.26, 3.27).

Occasionally, underreams (bells) have been omitted, either by inadvertence or as a consequence of poor supervision or inspection. This kind of defect seems improbable, but it has occurred (2.4). If such an omission is suspected, or if there is reason to suspect that an underream may have caved or collapsed before or during concrete placement, a conventional test boring can be drilled close to the shaft. It is difficult to be assured of the plumbness of a deep test boring, and, for this reason, failure to encounter the concrete of the underream may be inconclusive unless the actual alignment of the test hole is checked by suitable instrumentation (inclinometer, etc.).

A Practical Low-Cost Full-Scale Load Test. Full-scale load tests on completed piers have been very expensive, and sometimes so difficult as to be considered unfeasible. Sometimes, when owner or designer insists on load tests, the solution has been to substitute two (or more) smaller piers for one large one, thus getting each pier load down to a magnitude that can be duplicated (or perhaps doubled) in testing. The load tests are still expensive; and the multiple piers to replace a single one are an added expense. As an alternative solution, sometimes drilled pier foundations have been conservatively overdesigned, based on data from small-scale laboratory tests, from SPT or CPT data from the geotechnical investigation, or from small-scale in-situ field tests (torvane, pocket penetrometer, etc.), plus past experience with foundations in the same formations.

In 1984, Dr. J. O. Osterberg designed and tested a method of load-testing full-scale drilled piers which is relatively easy to apply, and relatively inexpensive, costing only a fraction of the cost of conventional load tests using either a reaction load or hold-down shafts and a heavy reaction beam (3.28). The method measures separately end-bearing load versus deflection, and sidewall shear (friction) versus deflection, at each load increment. This provides reliable data for design using both end bearing and sidewall shear in determining allowable pier loads. Essentially, the system consists of a flat pressure cell covering the entire bottom of the pier hole, a hydraulic pressure line from the ground surface to the pressure cell, and a rod and dial gages to measure the downward movement of the bottom and the upward movement of the shaft as pressure is applied in the pressure cell (see Fig. 3.17). A patent application for the device has been filed.

An advantage of this system is that, when the test is complete, the pressure cell can be grouted, and the test pier can be used as an element of the structure's foundation.

3.13.3. Repair of Defective Piers

In instances where a defect is located in the upper part of a pier, either above the water table or in a situation where the water level can be drawn down readily, the best and surest method of repair is to expose the defective concrete by excavating alongside the pier, perhaps by overdrilling the shaft with a modified core barrel. Concrete can then be removed and replaced with high-strength concrete such as prepacked concrete, or cementing com-

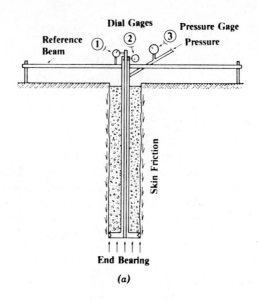

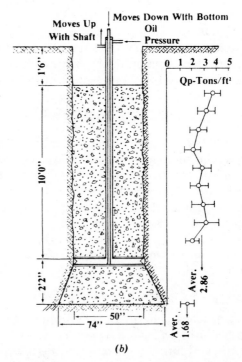

FIGURE 3.17. (a) Osterberg test of straight-shaft pier (schematic). (b) Osterberg test of belled pier; test set up to measure separately load capacity of belled base and shaft.

pounds such as epoxy. In some cases, dowels or additional reinforcement may have to be incorporated to compensate for uncertainties in bond at cold joints. Repairs of this type can be facilitated by the use of fiberglass or fabric forms especially designed for jacketing deteriorated marine piling. Where exposure of defects is not feasible, the installation of additional reinforcement by drilling and grouting techniques is most often used.

Reinforcement. Short defective lengths of a pier can be repaired by drilling core holes (with diameters at least 1.5 in. larger than the reinforcing bar) down through the defective zone. Steel bars of sufficient cross section to carry the entire vertical load and long enough to pass through the defective zone are then grouted in place, with care to prevent air blocks during grouting. Where appropriate, the pressure grouting will not only serve to bond the bars, but may fill communicating voids in the defective zone. This method of correction is expensive and requires meticulous work, but for deep piers it will usually cost substantially less than either complete replacement or side excavation and manual repairs.

Pressure Grouting. Grouting of defective concrete, like diamond coring, requires the best equipment and techniques. Moreover, if it is to be successful, special attention must be given to the choice of a suitable grout material. For the large open voids, an ordinary neat cement grout or a sand–cement grout is appropriate. If penetration into uncemented concrete aggregate in a segregated zone is required, a less viscous grout is needed. Grout containing superfine cement (3.29) has the lowest viscosity of the cement grouts and is recommended for permeation of marginally groutable zones. Colloidal types of cement grout dispersed by a high-speed machine, such as the "Colcrete" mixer (3.30) will also penetrate gravel and coarse-sand-sized aggregate, whereas ordinary cement grout may not. The reliability of the verification techniques described in Subsection 3.13.2 to verify the integrity of a grout repair is difficult to evaluate. Consequently, it is recommended that both coring and geophysical verification methods be applied.

Replacement Piers. Occasionally, the least expensive way to repair (or replace) a defective pier would be to construct new drilled piers to support the column load on a short girder spanning the defective pier. This technique is usually possible only for correcting defects discovered before the super-structure has been erected. Correction at a later stage of construction, when there is insufficient overhead clearance for conventional pier construction, is much more difficult. Under these circumstances, special drilling or manual excavation methods are required (including temporary column supports) to replace a defective pier. These underpinning procedures are extremely expensive and time consuming.

3.14. Housekeeping—Job Organization

A good, smooth-running drilled pier job requires more attention to some surface details than does, for example, a pile foundation job of the same size.

Surface drainage is especially important in pier-drilling operations. Drainage should always be away from a hole being drilled. The surface casing (if any) should be high enough to prevent fall-in of spoil or drainage of surface water into the hole, and a slight mounding of spoil around the hole may be allowed to keep drainage away from the hole until concrete has been placed. Good surface drainage over the entire site will facilitate the work in bad weather, especially where clayey soils are involved, and care should be taken to prevent ponding whenever the job is shut down.

Spoil from the drilling must be disposed of. Clayey spoil spilled in the work area may seriously impede operations; and if carelessly spilled on public thoroughfares while being hauled away, it can even cause a job to be shut down. Loose spoil left too close to an open hole may fall in, impeding the work, endangering workers in the hole, or contaminating concrete when the casing is pulled during or after concreting.

Open holes must be covered securely whenever the job is shut down, not only as a safety measure, but also to prevent entry of rain or snow. An open hole should *never* be left uncovered overnight.

Unlike most piles, drilled piers require careful scheduling of deliveries. In most circumstances, concrete placement must follow promptly on completion of drilling. This requires precision timing of concrete delivery, of delivery and placement of reinforcing cages, and of cranes for handling cages and for placing and pulling casing.

3.15. Drilled Piers for Retaining Walls

The application of drilled pier techniques is becoming increasingly common in the construction of retaining walls for excavations (3.31, 3.32, 3.33, 3.34, 3.35). An outstanding advantage of this type of retaining wall is that it can be designed to be watertight and to take full earth and groundwater pressure. Such walls can, under many circumstances, be cantilevered or designed with tiebacks, thus requiring no bracing within the excavation, and therefore offering no impediment to construction operations within the excavation.

This type of construction is often done in a sequence as follows:

1. A series of pier holes are drilled around the periphery of the proposed excavation, to a depth well below the proposed bottom of excavation and spaced less than two diameters apart, center to center. Care must be taken in drilling these holes, so that an open hole, or concrete that has been placed but has not yet set, is not disturbed by the drilling of an adjacent hole. It is often necessary to drill and concrete alternate holes first, then return and complete the series after the concrete of the first group has set.

These holes are concreted and reinforced as piers, designed to withstand external soil and groundwater pressure on the excavation wall when the excavation is complete. The design may involve taking the side pressure either by cantilever action or by the installation of post-tensioned earth or rock anchors, as illustrated in Fig. 3.18. In some designs, some or all of these piers are also used as the outer foundation piers for the structure. The foundation piers may have to be drilled deeper than the others to provide adequate bearing for the structural loads, in addition to whatever lateral resistance is needed for retaining-wall action.

2. When the concrete in the first series of piers has set up, a second series of holes is drilled, intermediate between the piers of the first series and displaced enough toward the outside of the excavation that each hole touches the concrete of the two adjacent piers (hence the name "tangent wall" that is sometimes used). These holes are cleaned carefully to make sure the concrete of the adjacent piers is exposed, then they are filled with concrete, completing the wall. The second series of piers may go to the full depth of the first series or may be stopped shortly below basement-floor level, depending on soil or rock conditions. They may be reinforced to take some of the thrust of soil and groundwater pressures or not, depending on the design and reinforcing of the first series.

3. When the wall has been completed as described above, the excavation is started. If lateral earth pressures are to be taken by an earth anchor

FIGURE 3.18. Drilled shaft wall with full cantilever above top of rock (excavation is 65–70 ft deep), and tied back with tendons at base of piers. (Photo courtesy of Mason-Johnson Associates, Inc., Dallas, Texas.)

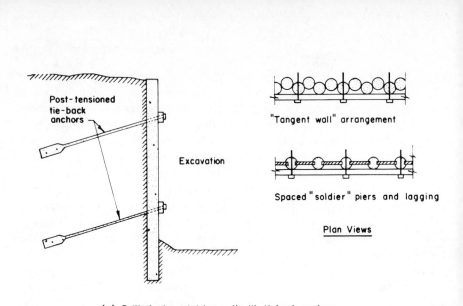

"Tangent wall" arrangement

Spaced "soldier" piers and lagging

Plan Views

(a) Drilled pier retaining wall with tieback anchors

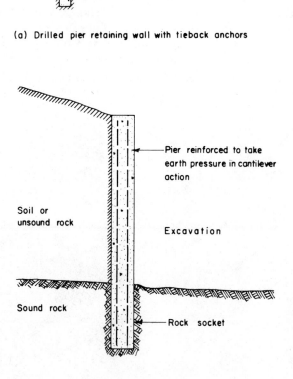

(b) Cantilever type drilled = pier retaining wall

FIGURE 3.19. (*a*) and (*b*) Types of drilled pier retaining walls.

system instead of by cantilever action in the piers, excavation is interrupted as soon as it is deep enough to permit drilling of the anchor holes. The anchor holes are drilled and anchors installed, the anchor rods are post-tensioned, and excavation is resumed down to the next level of anchors (if any) (3.36). Anchor holes are usually drilled with a special machine, using a long continuous-flight auger. They may be horizontal or sloping, depending on the soil and rock formations to be penetrated. When rock anchorage cannot be obtained, sometimes conventional pier holes with underreams are put in, at a suitable batter, with pier-drilling machines.

Figure 3.19 shows a drilled pier retaining wall cantilevered above top of rock and tied back at base of piers.

4. In many cases, the drilled pier retaining wall is combined with the basement wall by casting a concrete facing against the inside of the retaining wall. In other cases, a separate wall is constructed, the space between the piers and the inner wall being used as a drainage gallery to take care of any seepage through the outside wall.

Many variations of these designs have been worked out to fit the requirements of the structure and the local soil or rock conditions. Sometimes the spacing between the holes of the first series is nearly two diameters, and the holes of the second series are the same diameter as the first; or the

FIGURE 3.20.   Watson tieback drill for anchor holes.

first series may be closer together, with the second series much smaller. Sometimes only one series of piers is installed, with a spacing of three or more diameters; grooves for insertion of precut lagging are formed by positioning metal forms, or foamed plastic strips, on the sides of the pier hole before concrete is placed; and lagging is added as the excavation progresses (3.31).

Figure 3.20 shows a specialized drilling machine for drilling small-diameter anchor holes in soil or rock.

Drilled piers have also been used in conjunction with mud slurry to construct retaining walls. The piers are completed at a spacing of three to five diameters. The soil between piers is then excavated, the removed material being replaced by slurry as excavation proceeds. Generally, short trench sections are completed alternately; reinforcing is placed according to the design; and the slurry is displaced by concrete, pumped in at the bottom of the trench. When the concrete in the first set of trench sections has set adequately, the intervening trench sections are excavated and the wall is completed. Tiebacks are generally used to take lateral loads (3.35).

References

3.1   Sheikh, S., 45 Degree Underreams Undergo Further Testing, (Project in the Structural Capacity of 45° Underreams), *Foundation Drilling*, Dec./Jan. 1985; also earlier article, same publication, Aug. 1983.

3.2   Rigid Casings Speed Wet Sand Bores, *CM&E*,* March 1964 (4 pages—reprint from Case International, Chicago). Thirty-three piers to support American Dental Association Headquarters Building in Chicago. Nine piers were 8 ft diameter, with underreams to 24 ft diameter, for column load of 8,700,000 lb. Borings went through saturated sand, soft and stiff clay, by patented "Case" method, forming slurry of sand and bentonite through 30-ft sand layer; 10-ft-diameter casing set through slurry-filled hole and sealed into clay by forcing down several feet; then slurry was removed and 8-ft-diameter hole drilled neat through clay, in the dry. Smaller steel liner was set temporarily through clay while 24-ft bell was mined by hand, using pneumatic spades and paving breakers.

3.3   Biggest Mechanically Built Caisson, *CM&E* (2 pages—reprint from Case International, Chicago). Describes same project and procedures as Ref. 4.5.

3.4   Mohan, D., Jain, G. S., and Sharma, D. Bearing Capacity of Multiple Underreamed Bored Piles, *Proc. 3rd Asian Reg. Conf. SM & FE, Haifa*, pp. 103–106, 1967. Report of experiments indicating that load settlement of underreamed piers in sandy and clayey soils is substantially improved if two or more underreamed bulbs are provided, and that an increase in bearing capacity of about 50 percent can be obtained with only 30 to 40 percent of the concrete required for a straight-walled pier of the same diameter.

---

* *CM&E = Construction Methods and Equipment.*

3.5 New Hospital Doubles Up Over Older One, *ENR*, Nov. 13, 1969. Foundations for a 13-story tower straddling the existing Kaiser Foundation Health Plan Hospital in Oakland, Calif., were drilled piers, 38 in. in diameter and 120 ft deep. The lower 50 ft of each pier hole was grooved in 1-ft rings on 3.5-ft centers to assure development of sidewall shear in the foundation soil.

3.6 Pells, P. J. N., Rowe, R. K., and Turner, R. M., 1980. An Experimental Investigation into Side Shear for Socketed Piles in Sandstone, Intl. Conf. on Structural Foundations on Rock, Sydney, Australia.

3.7 Drill Rotates Caisson Liners into Bedrock, *CM&E*, August 1964, pp. 96–99. Pier foundations for a Chicago 30-story building were drilled through clay, hardpan, silt, and loose rock to rest on bedrock at about 100-ft depth. Bottom liners of fabricated steel were twisted into the rock to form a seal for dewatering.

3.8 Moore, W. W., Foundation Design—the Golden Gateway, *Civil Eng.*, January 1964, pp. 33–35. Heavy structures in San Francisco's Golden Gateway project are founded on skin-friction piers in weathered sandstone and shale. Piers pass through variable depths of fill and bay mud. Design stresses of 10,000 psf were used, based on a load test in which support was by sidewall shear only. Holes were drilled 54 in. diameter in soil, 48 in. diameter in rock, with casing sealed into rock. Drilling was done in mud slurry made from soil being drilled; slurry was thinned to "consistency of heavy cream or buttermilk"; density was held to not more than 85 pcf. Smear on rock sockets was cleaned by scraper teeth on outside of cleanout bucket, sometimes by water jet.

3.9 Bietz, J., and Waite, W. Solids Control for Slurry Displaced Drilled Shafts and Diaphragm Walls—Settlement or Mechanical Separation?, *Foundation Drilling*, March–April 1985. European practice in control testing and cleaning of drilling slurries is described.

3.10 McKenzie, I. L. Comparison of Pile Load Tests for Piles Installed Dry and Under Bentonite Slurry. Symp. "In-situ Testing for Design Parameters," Melbourne, 5 Nov. 1975.

3.11 Fearenside, G. R., and Cooke, R. W. The Skin Friction of Bored Piles Formed in Clay Under Bentonite, CIRA Report 77 (1978). Load tests of bored piles in London clay indicate no reduction in sidewall support due to placement under bentonite slurry.

3.12 Gupton, C. P., and Logan, T. J. Design Guidelines for Drilled Shafts in the Weak Rocks of South Florida, Paper presented at the 1984 South Florida Annual ASCE Meeting, September 1984.

3.13 Smith, M. L., Sheath of Sand Salvages Steel Caisson Liners, *CM&E*, October 1964, pp. 118–123. An outer (temporary) casing is set through water and sand and sealed into clay, an inner temporary casing is centered in the first one, and the annular space is filled with sand. The boring is then extended into the clay, the underream is formed, and steel and concrete are placed in the usual way. As soon as the concrete is in place, the inner casing is pulled carefully, leaving the concrete retained by a smooth layer of sand. When the concrete has set, the outer casing is pulled and the sand sloughs off, leaving the smooth column of concrete. Illustrated.

3.14 Sheikh, S. A., O'Neill, M. W., and Mehrazarin, M. A. Application of Expansive Cement in Drilled Shafts, 170-page report available from ADSC, Box 280379, Dallas, Texas 75228.

3.15   Bad Caissons on 500 North Michigan Avenue, *ENR*, Sept. 29, 1966, p. 14. Trouble in drilled piers for 25-story office building is attributed to fog in the hole during cold-weather concreting, which prevented proper inspection; also, no casing was used.

3.16   Baker, C. N., and Kahn, F. Caisson Construction Problems and Methods of Correction, *ASCE Preprint* 1030, October 1969. A good paper, going into some detail about types of flaws, the conditions which can cause them, and corrective measures.

3.17   "Investigation of the Concrete Free Fall Method of Placing High Strength Concrete in Deep Caisson Foundation," report prepared by Soil Testing Services, Inc., for Case Foundation Co., April 1960.

3.18   Randall, Frank, Design of Concrete Foundation Piers, *Portland Cement Assoc. Concrete Rept.* XS 6830. This paper recommends low-slump gap-graded concrete for deep piers. This recommendation is acceptable for straight-shaft piers without reinforcing cages of large underreams. It is wrong if the pier has either of these features.

3.19   Concrete Trips on Aluminum Pipe, *ENR*, Dec. 18, 1969, p. 57. Aluminum pipe declared unsuitable for tremie or pumped concrete; BPR has advised field offices and state highway departments to this effect. Tests produced hydrogen foaming; concrete expanded; loss of strength at 14 days was 9 to 14 percent.

3.20   Bernal, J. P., and Reese, L. C. Drilled Shafts and Lateral Pressure of Fresh Concrete: New Research Findings, Foundation Drilling, May 1984, p. 11.

3.21   Hearne, T. M. et al. Drilled Shaft Integrity by Wave Propagation Methods, *Proc. ASCE, Vol. 107*, GT10, 1981.

3.22   Griffiths, D. H., and King, R. F. Applied Geophysics for Engineers and Geologists, Pergamon Press, 1965.

3.23   Preiss, K., and Caiserman, A. Non-destructive Integrity Testing of Bored Piles by Gamma Ray Scattering, *Ground Engineering* 8 (1975) No. 3. Gamma ray backscattering tests made in tubes and core holes located all faults that were subsequently located by coring or excavation, and many that were not located by coring.

3.24   Trantina, J. A., and Cluff, L. S. *NX* Bore Hole Camera, *Symp. Soil Exploration*, ASTM STP 382, 1963, pp. 108–120.

3.25   The *NX* Bore Hole Camera, *ENR*, June 25, 1953.

3.26   Wilson, S. D. The Use of Slope Measuring Device to Determine Movements in Earth Masses, *Symp. Field Testing Soils*, ASTM STP 322, 1962, pp. 187–198.

3.27   Digitilt, *Indicator*, vol. 1, no. 2-3, pp. 6–7.

3.28   Osterberg, J. O. A New Simplified Method for Load Testing Drilled Shafts, *Foundation Drilling*, August 1984.

3.29   Shimoda, S., and Ohmori, H. Ultra Fine Grouting Material, *Proc. Conf. on Grouting in Geotechnical Engineering*, ASCE, Edited by Wallace Hayward Baker, New Orleans, La., Feb. 1982.

3.30   Champion, S., and Davies, L. T. "Grouted Concrete Construction," *Reinforced Concrete Assoc. Meeting, London*, Feb. 12, 1958. This paper deals with grouted concrete made by the "Colcrete" process.

3.31   Subway Caisson Walls Trim Time, *CM&E*, September 1970, pp. 84–86. Wall constructed by piers 3 ft diameter, casing 42 in. diameter; rebar cages were fabricated with slot for Styrofoam filler to be removed after concrete hardened, making key for 8-in.-thick, 2-ft-long diaphragm wall section. Every fourth pier had 5-ft 30° bell at bottom. Sonotube was used to position and center cage and was pulled when concrete was placed. Shaft at wall level was 3 ft diameter, 42 in. in foundation below wall. Drilled with Calweld rig on Lima 30-T crawler crane.

3.32   Engstrom, H., and Oates, R. Drilled Pier Cofferdam for a Building, *Civil Eng.*, April 1967, pp. 44–45. A three-story basement below groundwater, in an area where drawdown could not be allowed, was constructed inside a watertight cofferdam constructed of "tangent" drilled piers. One hundred thirty-four holes 30 in. in diameter were drilled at 3-ft centers and concreted into rock and held at the top by shoring, then intermediate 15-in.-diameter piers were drilled outside and tangent to the first series. A 2-ft-thick concrete floor slab was tied down with anchors drilled into bedrock to resist hydrostatic uplift.

3.33   Andrews, G. H., and Klasell, J. A. Cylinder Pile Retaining Wall, *Natl. Res. Council Publ.* 1240, 1964.

3.34   Bulley, W. A. "Cylinder Pile Retaining Wall Construction—Seattle Freeway," *Roads and Streets Conf., Seattle, Wash.*, January 27, 1965, 11 pp. 2,630 unusual cylinder piles (drilled piers) were used in 4 miles of retaining wall, part of which was used to stabilize a moving landslide area. Hillside cuts were as much as 100 ft deep, in unstable and potentially unstable soil. Holes of 5.5, 8.3 and 10 ft diameter (1.5, 2.5, and 3 m), were drilled to depths of 33 to 120 ft (10 to 36 m) below ground. No anchors were used. The piles acted as cantilever beams, transferring the load to stable soil (dense glacial till). Welded I-beams were used as core steel, the largest having webs 100 in. (2.54 m) in depth and flanges as large as 30 in. × 3.5 in. (76 cm × 8.9 cm).

3.35   Andrews, G. H., Squier, L. R. and Klasell, J. A. Cylinder Pile Retaining Walls, *ASCE Preprint* 295, January 1966, 46 pp. A detailed discussion of the Seattle Freeway retaining wall: soil and geologic conditions, design, construction, and performance.

3.36   White, Robert E. Pretest Tiebacks and Drilled-in Caissons, *Civil Eng.*, April 1963 (3 pages—reprint from Spencer, White & Prentis, New York). Retaining wall for an excavation formed by vertical drilled-in caissons at wide spacing, held by 45° tiebacks drilled and grouted into rock and then posttensioned (Freyssinet system). Tension in tiebacks is 150 kips.

Bibliography

The following articles and papers, while not keyed to specific parts of this chapter, are of interest and value:

Drilled Caissons Cut Underpinning Cost, *ENR*, Sept. 3, 1964, pp. 36 and 41. A description and drawing illustrating underpinning of an existing building while a 20-ft basement excavation was being made adjacent to it. Grade beams spanned from rock located under the building to pairs of 18-in.-diameter piers drilled to deeper rock at the location of the proposed basement wall.

Houston's Distinctive Underpass Is Built in and on the Ground; Dug Out, *ENR*, July 19, 1962. Retaining walls are constructed below the water table as closely spaced drilled piers, then excavation is made, and finally a concrete curtain wall is constructed in front of the pier wall and anchored to the piers. Sand wells behind the wall are used to relieve hydrostatic pressures.

Big Down-the-hole Rig Drills Caisson Shafts, *CM&E*, December 1963, pp. 88–89 (the Magnum drill). Description of a pier drill with seven pneumatically operated hammer drills mounted on a rotary drill assembly, operated by a crane-mounted drilling machine. The drill assembly comes in 15, 18, 24 and 30 in. diameter; it is claimed to make 15 to 30 ft of penetration per hour. In a St. Louis job, a 24-in. unit drilled 58 piers 15 ft into limestone in 13 eight-hr days. Long-buried Tiebacks Leave Cofferdam Unobstructed, *CM&E*, May 1966, pp. 130–135.

Gauntt, C. G., Marina City—Foundations, *Civil Eng.*, December 1962, pp. 61–63. Sixty-six-in.-diameter reinforced caissons for towers of Chicago's Marina City, installed by "Case" method, to average depth 110 ft, through bouldery glacial till to bear on sound limestone. Groundwater sealed off by temporary casings; thin steel liners used for forms for shafts, and casings were salvaged and reused.

Horvath, R. G., Kenny, T. C., and Cozicki, P. Methods of Improving the Performance of Piers in Weak Rock, *Canadian Geotechnical Journal*, Vol. 20, No. 4, 1983.

Pearson, J., Arizona Bridge Project, Foundation Drilling, September, 1984.

Specification for Cast-in-place Piles Formed Under Bentonite Suspension, *Ground Engineering* 8 (1975) No. 2.

Construction Accuracy Essential for London Jumbo Piles, *Ground Engineering*, 7 (1974) No. 6. Describes instrumentation and techniques for controlling location of bored piles within 75 mm at surface and 25 mm at depth of 22.5 m.

"REVERT, a self-destroying fluid additive for use in wells and test holes," information bulletin from Johnson Division, Universal Oil Products Company, St. Paul, Minn.; also several reprints of articles on "REVERT" from the *Johnson Drillers Journal* (same source).

Kapp, M. S. Slurry Trench Construction for Basement Wall of World Trade Center, *Civil Eng.*, April 1969, pp. 36–40.

Custom-built Reverse Circulation Drill Rig Used in Constructing Offshore Platform, *Western Construct.*, April 1964, 2 pp.

Big Bell-bottom Caissons Will Support Detroit's Second Civic Center Building, *ENR*, Dec. 6, 1951, pp. 44–46.

Detroit Builds First Unit of Civic Center, *ENR*, Sept. 30, 1948, pp. 68–69.

Giant Augers Sink 15-ft-dia Missile Shafts, *CM&E*, May 1963, 3 pp.

Etheridge, D. C. Drill Digs Deep for Bridge Piers through Dam Abutment, *CM&E*, May 1970.

Caisson Trouble Hits Chicago, *ENR*, Sept. 29, 1966.

Scraper Teams and Giant Augers Pace Construction at Wyoming Missile Base Project, *Western Construct.*, July 1963, 4 pp.

St. Louis Floodwall Doweled to Rock, *ENR*, Apr. 9, 1964, 2 pp.

Vertical Mole Scores Bullseye in Deep Shaft, *ENR*, Feb. 27, 1969, pp. 28–29.

Five Wells to Drain Wet Foundation, *Western Construct.*, March 1957, 2 pp.

4

Engineering Construction Review and Inspection

Engineering construction review and inspection practices vary widely by both organization and geographical area. In the United States legal requirements for inspection differ from city to city. Government regulations vary between various departments and bureaus. Customs have developed differently in different areas, and variation in geologic conditions has led to major differences in regional engineering construction review and inspection practices.

In this chapter the authors present their conception of a comprehensive engineering construction review and inspection program for drilled pier construction. This program represents what we believe should be followed, rather than what the present practices actually may be.

4.1. Responsibilities of the Engineer and the Contractor

It should be clearly understood that the phrase "engineering construction review and inspection" applies only to obtaining compliance with the "real" intent of the plans and specifications. This implies that every reasonable effort should be made to ensure that the materials used—and construction procedures, to the extent that they are spelled out or would adversely affect the intended quality of the end product—should comply with the plans and specifications.

Engineering construction review and inspection is not intended to imply that any direction is given to the contractor or subcontractors, or any of their employees, as to how they carry out their operations. The contractor's (and subcontractors') operations are strictly their own responsibility. This includes any provisions they do or do not undertake with respect to the safety of personnel on the job site. The contractor should be given every opportunity to exercise his ingenuity in developing methods that most efficiently will accomplish the requirements of the plans and specifications.

The engineers (and architects) have the responsibility of carefully and accurately developing plans and specifications that will provide the owner with a functional structure compatible with his financial capacity, and that also will protect the safety of the public. For structures involving drilled piers, the plans and specifications definitely should include a copy of the complete geotechnical report. While these documents normally leave field procedures to the ingenuity of the contractor when drilled piers are specified, certain procedures may properly be required to ensure the quality and adequacy of the end product. Obvious examples are the requirement for a dewatering system, or the specification of concrete placement by tremie or by pumping under certain circumstances.

4.2. Scope and Limitations of Construction Review and Inspection

Members of the design professions should perform some on-site review on all construction projects to ascertain that the construction is proceeding, in general, in conformance with the intent of the plans and specifications. In the authors' opinion it is essential to the owner's interest to require construction review and inspection on all drilled pier installations. The extent of the services should be governed by the importance of the construction, by the complexity of the subsurface conditions at the site, and by local code requirements. In areas of known uniform geologic conditions, generally favorable for drilled pier construction, observation of the drilling operations and inspection of the completed pier holes before concreting may be adequate if light structures are involved. Even under such favorable circumstances, however, the person performing the inspection should be knowledgeable with respect to possible problems. The underground is never completely predictable, and the construction review and inspection should never be superficial or casual.

Continuous construction review and inspection are imperative on projects where the owner's investment is large, where heavy loads or special structural requirements are involved, where geologic conditions are complex, or where other unfavorable conditions could be encountered.

Where the owner and the contractor (or the designer and the contractor) are one and the same, an employee of that organization, performing construction review, can be subjected to pressures that could lead to an unsound decision in order to save money in construction. A man not subject to such pressures is in a position to make sounder decisions. The authors feel that it is generally to the owner's interest to have construction review and inspection done by an independent agency.

4.2.1. Definitions

Unfortunately, in the past the terms "construction review," "supervision," and "inspection" often have not been clearly defined in specifications, contracts, or other documents. In such cases there often is a tacit assumption that the meanings of these terms are clear and unarguable. The record of many lawsuits testifies to the fact that this is not true. Several decisions have been rendered by the courts holding that "supervision" in particular extends to responsibility for the adequacy of temporary bracing or scaffolding, and even to the manner in which contractors' employees operate equipment. The use of the term "supervision" to designate the service described above as "construction review" should be avoided. "Construction review" and "inspection" require careful definition in the contract documents, with clear statements of both scope and limitations. Moreover, the definition must be tailored to the task at hand.

In recent years professional societies representing architects and engineers have made rapid strides in developing contract documents which more clearly define "construction review" and "inspection." These documents are not established to the degree, however, that their definitions will automatically be accepted in a court of law. Also, it is difficult, in such standard documents, to describe accurately such activities when drilled piers are involved. Therefore, these activities should be defined carefully in the contract documents for drilled pier projects.

The terms "construction review" and "inspection," as used in this book, refer to the activities and responsibilities of the individual, agency, or firm who represents the owner's interest in determining that the materials used and the work done are in accordance with the intent of the plans and specifications (4.1). The engineer assigned to this duty, covering the foundation phase of a project, is often referred to as the "geotechnical engineer" or the "field engineer." "Inspection" is a part of his responsibility, and he may have project inspectors reporting to him and being supervised by him.

On large projects, the owner is often represented by a "resident engineer," who remains permanently on the project and manages it.

4.2.2. Obligations

The engineer performing construction review has a direct contractual obligation to his client, who normally (and ideally) is the owner, but who may be another engineer, an architect, or an engineer-constructor. Also, he has an implied obligation to the public, as a consequence of his registration or license as an engineer and his approval of the plans and specifications. The engineer should cooperate with the contractor to avoid any unnecessary delays. However, such cooperation must not conflict with his obligations as described above.

In foundation work, or in any construction work on, in, or of soil or rock, the occurrence of so-called "changed conditions," which more properly should be termed *unexpected* conditions, is not unusual. The duty of the field engineer under such circumstances normally requires him to make field decisions. These may cause him to recommend variations in construction operations or techniques, materials, or even structural elements, which were not anticipated in the contract documents but which are necessary to carry out the *intent* of these documents.

The field engineer's primary obligation is to see that the work conforms with the intent of the plans and specifications; however, he also has an obligation to use his best efforts to keep the job moving smoothly, to avoid any interference with the construction schedule. During the work he provides advice and recommendations to the owner or his representative. If he finds defects in construction or materials, it is his duty to notify the owner's representative so that the error may be corrected; and as part of that duty, he keeps the contractor appraised of his advisories to the owner. A cooperative relationship between the contractor and the field engineer is always desirable, but the engineer's primary responsibility is to his client.

When the work is complete, the engineer renders a professional opinion as to whether or not it conforms to the full intent of plans and specifications.

In many instances, the field engineer will be required to submit a report to the city building inspection department (or other authority) stating his opinion regarding the conformance of the completed work to the plans and specifications. Such a report implies a judgment of the suitability of the design for the proposed use, a function that may not actually be included in the duties of the field engineer. The owner, however, will expect a final report that will be accepted by the authorities as the basis for issuing a certificate of occupancy. Thus, it is imperative that the field engineer should satisfy himself in advance regarding the adequacy of the plans and specifications, as well as what opinion or certificate he will be asked to sign. His construction review and inspection then should be thorough enough to enable him to provide a report that will meet these requirements.

As construction progresses, the field engineer should immediately advise the owner's representative of the use of any procedures or materials that do not comply with the intent of the plans and specifications. If corrective action is not taken, he then must advise the owner's representative—in writing—that he will not be able to render an opinion that the completed work complies with the contract documents. If corrective action still is not taken, the engineer should sever his connection with the project, immediately and in writing. This action is necessary to protect both the engineer and the owner. Although the owner may not approve this action, it does protect him, not only as to the adequacy of his structure, but more importantly with respect to his obligation to protect the safety of the public.

4.2.3.   Limitations of Construction Review and Inspection

It has been emphasized that the terms "construction review" and "inspection" do not imply the exercise of any supervisory authority over the contractor, including his employees, or any responsibility for their actions or lack of action. It is the responsibility of the contractor—not the field engineer—to run the job. The field engineer should not direct the contractor in the performance of his work, nor ask him to make changes outside the scope of the contract, unless the owner's representative has issued him written instructions to do so. Even though—in his judgment—the field engineer sees work being performed incorrectly, he must resist the temptation to "take charge" and direct the workmen to correct it. To yield to such temptation—no matter how obvious the error is, or how easily corrected—would put the field engineer in the position of sharing with the contractor the responsibility for the performance of the work. In the event of a dispute, the record of such an action would permit the contractor to claim that he had relied on the advice of the field engineer in the carrying out of his (the contractor's) operations throughout the rest of the job. If changes become necessary or desirable, the field engineer can then, under the conditions described above, make *recommendations* for corrective action. Any *directives* to the contractor should come from the owner.

When drilled piers are involved on a project, the field engineer frequently may find it necessary to recommend changes in operational techniques or in the quality or quantity of materials being used, or even to request additional work outside the scope of the original contract. Therefore it is imperative that the scope of his authority with respect to construction review and inspection be clearly described in all contracts, documents, or agreements that could affect the engineer (or architect) involved. Such documents include the contract between the owner and the contractor. The engineer (or architect) performing construction review should require, in this document, protection with respect to "injured workman" occurrences (discussed in Chapter 7).

4.3.   Duties and Qualifications of Geotechnical Engineer and Foundation Inspector

Because of the specialized nature of the problems and the frequent need for continuous expert attention, engineering construction review involving drilled piers is often delegated to a specialist (person or firm), generally designated as the "soil engineer," "soil and foundation engineer," or "geotechnical engineer." The authors will refer to such a specialist as the "geotechnical engineer," a term that embraces the fields of soil and rock mechanics, groundwater hydrology, and engineering geology as they apply to foundation and earthwork problems. The duties of a geotechnical engineer, in carrying out construction review (and inspection) with respect to the geotechnical phases of a project, include the following functions (which must be at the request of the owner's representative, who frequently is the architect on building projects involving drilled piers):

1. Observation of construction procedures, noting deviations from the true intent of the plans and specifications.
2. Advising and conferring with the owner, architect, structural engineer, and contractor.
3. Identifying and preparing needed specification revisions when authorized by the owner's representative.
4. Acting as "interpreter" to be sure that owner, architect, structural engineer, and contractor understand each other in geotechnical matters.
5. Responsibility for foundation construction review, inspection, and reporting.

The *inspection* duties involved in construction review which are discussed here may be performed by a geotechnical engineer; or they may be performed by a technician-inspector who reports to the owner, to the project engineer or architect, or to a geotechnical engineer. In the discussion that follows, the term "foundation inspector" will be used to designate the person performing this function, regardless of whether he is a technician or an engineer. Figure 4.1 shows a typical arrangement of lines of authority, from owner down to inspector.

The foundation inspector on a drilled pier operation is required to make field decisions on matters much more complex (and less certain) than an inspector on typical concrete or steel construction, for example. If the inspector is a technician (or an engineer who is not registered or licensed), he acts as the "eyes" of the geotechnical engineer, whose duties include the exercise of engineering judgment. The inspector observes, makes reports, applies criteria established for him by the geotechnical engineer, and limits his field decisions to the application of these criteria. He refers matters requiring engineering decisions to the geotechnical engineer. But because the need for field decisions of an engineering grade is common in drilled pier work, lines of communication between the foundation inspector and the geotechnical engineer must be well established, and the foundation inspector must be able to get prompt decisions whenever questions requiring engineering judgment arise. Delay in obtaining such decisions risks impeding the contractor's operations. The making of such decisions by personnel not technically or legally qualified could involve the responsible agencies in dangerous and unnecessary liabilities.

It will be assumed, in the discussion that follows, that the person acting as foundation inspector will have the training and ability to perform the duties described. A foundation inspector is qualified by training and experience, not by appointment or self-proclamation. The degree of training

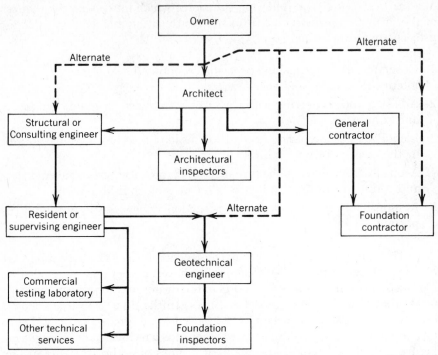

FIGURE 4.1. Lines of authority in the field organization of a typical building construction job.

and experience required depends on the size and importance of the structure and its relation to public safety, as well as on the complexity of subsurface conditions. A man may be quite competent to act as inspector of drilled pier construction for a one-story warehouse located on relatively good ground, but lack the training needed to perform the same functions for pier holes in weathered or decayed rock or for a bridge or a high-rise apartment building.

Some jurisdictions register or license inspectors (e.g., San Diego, California). The organization supplying or engaging foundation inspectors should make sure of compliance with all relevant laws and ordinances.

For a large drilled pier project that requires continuous geotechnical engineering construction review, the usual arrangement is to use foundation inspectors who report to the geotechnical engineer (see Fig. 4.1). Their duties usually are (1) to make observations and records and report to the geotechnical engineer for decisions, (2) to identify bearing strata, and (3) to observe the contractor's operations for compliance with plans and specifications and notify the geotechnical engineer of any deviations. In this case, field decisions will be made by the geotechnical engineer, or, in his absence, by the foundation inspector, acting on the basis of specific criteria and instructions given him by the geotechnical engineer. It is obvious that this action by an inspector requires a special degree of technical competence. The authors have seen many instances where drilled pier costs ran far over contract or budget costs because the "inspectors" did not recognize suitable bearing material and did not approve stopping the drilling when competent soil had been reached.

Some construction contracts state that the field engineer or the inspector has the authority to require the contractor to comply with all applicable safety standards and codes. Neither the geotechnical engineer nor the foundation inspector should assume this duty unless directed to do so in writing, for to do so is to incur (or share) a liability that properly belongs elsewhere. However, if the construction contract names the geotechnical engineer (or soil engineer, or soil and foundation engineer, or even "the inspector" without designating specifically the foundation inspector), the geotechnical engineer would be well advised to obtain written instructions covering this matter. With this mention in the construction contract, and lacking specific instructions *not* to assume this duty, he should assume that a judge or jury would interpret the construction contract as having charged him with responsibility for the contractor's observance of safety rules—and that he will therefore share the contractor's liability in the event of a construction accident, even though his own contract or instructions may not mention the subject (4.2). In this case he is incurring a liability that he is probably not being adequately compensated for—a condition that he should not accept (4.3).

In the author's opinion, the obligations and qualifications that go with specialist positions for construction review and inspection are as follows:

Geotechnical Engineer

Obligations. The general duties and responsibilities of the geotechnical engineer with respect to his assignments are to represent the interests of his client, and to use his knowledge and expertise, acting in an advisory capacity, to keep his client apprised of any construction errors which could lead to increased costs, or to later structural distress with consequent correction, increased maintenance costs, or danger to the public.

The geotechnical engineer's construction review services on a drilled pier operation are confined to the drilling, the placement of reinforcing, and the concreting of the drilled pier foundations. He provides the inspection services which comprise an important part of his duties; makes field decisions when necessity arises; keeps records and makes regular, periodic reports; and provides technical liaison between owner, designer, and contractor.

Note that the geotechnical engineer does not normally perform surveying functions. He does not lay out or independently check the pier locations; and he does not determine or check their as-built positions, unless this work has been assigned and he has been provided with adequate reference points to facilitate measurements. Proper survey control must be established before the geotechnical engineer can proceed with such work, if assigned.

When drilled pier construction is complete, the geotechnical engineer submits a report stating his opinion of the conformance of the construction to the plans and specifications. In some cities, he may be asked to certify as to compliance. This does not come within his duties, and he should not allow himself to be induced to "certify."

Qualifications. The geotechnical engineer should be a civil engineer or engineering geologist, trained and experienced in the art of soil and foundation engineering. Training in or ready access to expertise in rock mechanics is needed on some projects where loads are heavy and rock conditions present problems, such as pier holes in cavernous limestone or dolomite. Training and experience in engineering geology are very useful in this field, and of great importance on some jobs. Primary training in engineering geology, with supplemental training and experience in foundation engineering, often provides excellent qualifications for this position. (To avoid conflict with the engineering registration laws, a geologist in this position should not be designated as an engineer unless he is licensed as an engineer.) An engineer and geologist, working as a team, often provide very good coverage for geotechnical problems.

Equally important to his understanding of geotechnology, the engineer must be well versed in the drilled pier construction process and in the engineering characteristics of the materials used for pier construction. This qualification is acquired only by relevant experience.

Foundation Inspector

Obligations. The foundation inspector is expected to perform the following duties.

1. To be familiar with the plans and specifications.

2. To identify predesignated bearing strata when they are encountered in the hole.

3. To make measurements of hole depth, plumbness, and underream diameter if appropriate.

4. To inspect the hole for cleanness and freedom from excess water before concrete is placed.

5. To observe concrete and steel placement.

6. To note, record, and immediately advise the geotechnical engineer and the owner's representative of any practices used that might adversely affect the compliance of the finished pier with the intent of the plans and specifications.

7. To check the locations of the finished piers against designated locations (but only if directed to do so and if adequate reference points have been provided).

8. To keep daily and weekly records of all pier construction work, including all reports issued regarding unsatisfactory work or construction practices.

9. To record remedial measures taken and changes made in construction practice.

The foundation inspector may be required to sign a report of pay items completed by the contractor, at the end of each work day.

All significant communications to the geotechnical engineer should be in writing, with copies to all concerned.

McKaig (4.4) states:

> The inspector should always remember that the work may become the subject of litigation even though all parties try to avoid such a development. He should therefore maintain his records in such a manner that his testimony, if needed, will be amply documented.

Qualifications. The foundation inspector should be able to perform competently the functions listed above to the extent that they are required on the job. On jobs where there is no geotechnical engineer, the foundation inspector should have not only the ability to identify the assigned bearing strata, but also sufficient experience in foundation work to enable him to recognize unfavorable or unanticipated soil or rock conditions and improper construction practices.

In addition to the technical qualifications listed above, the foundation inspector and the geotechnical engineer should have the personal characteristics of integrity and intelligence. A most important qualification for either of these positions is the ability to "get along" with the contractor and his personnel while maintaining a position of objectivity.

4.4. Preconstruction Preparation

4.4.1. Study of Available Data

The foundation inspector should, before going on the job, familiarize himself with available data that may relate to his work. These will include the plans and specifications, the geotechnical report, if any, as well as pertinent local building code provisions. In the event that a complete geotechnical report is not available, and only the boring logs are reproduced on a page of the plans (a practice that should be discouraged), the inspector should study the logs and examine the soil and rock samples, particularly those from the assigned bearing stratum and the strata immediately above and below it. It is the owner's obligation to have these samples available for the inspector's use in the field. These samples should be preserved at their natural water content for this purpose. Many soil and rock materials look quite different after they have been allowed to dry. Rock cores should be surface-wet when examined, when appropriate, to make sure their appearance coincides with that of material as it is brought out by the pier-drilling auger.

4.4.2.   Determining Scope of Assignment

The scope of the foundation inspector's assignment varies as mentioned earlier. This scope should be defined in the construction contract, for the contractor's information; it should be defined in the agreement between the firm furnishing the inspector and the owner; it should be set forth clearly in the foundation inspector's instructions, which should be in writing; and—very important—these three documents must be in agreement with one another.

In some instances the pier design is based on support by sidewall shear alone, and only a cursory inspection of the bottom of the hole is required. However, the authors are aware of two such projects where shaft friction design values were not attainable and end bearing had to be invoked. In such instances, the foundation inspector will be required to enter the hole and examine the bottom closely, or to observe the drilling of proof-test holes in the bottom. In some cases there will be a concrete inspector on the job to make test cylinders and slump tests; on other jobs the foundation inspection may include these duties (but only if the inspectors have been properly trained). When the hole is to be stabilized by the use of slurry, the foundation inspector may be required to check the density and viscosity of the slurry as drilling proceeds. Other variations in the duties of the foundation inspector are possible, and everyone involved should know what he is not expected to do as well as what his assigned duties are. Often this determination requires a preconstruction conference including the owner, the designer, the resident engineer (if any), the geotechnical engineer, and the foundation inspectors.

4.4.3.   Site Visit and Meeting with Contractor

Before a drilled pier operation is started, it is often helpful for the geotechnical engineer and the foundation inspector to visit the site; to discuss with the contractor, the owner's representative, and others involved the inspection duties and their possible effect on the contractor's operations; and to make sure that the contractor will have at the site any equipment that he must furnish to enable the foundation inspector to perform his assigned tasks. As discussed in Chapter 6, the authors prefer to see specifications and contracts so written that a contractor has flexibility in equipment and methods, and is not handicapped by restrictions on either. But there are instances where specific local experience has shown that it is to the owner's advantage to specify a minimum acceptable torque for the drilling machine, for example. The foundation inspector should be aware of any such requirements or limitations on equipment in the plans and specifications. He should report to the geotechnical engineer and have the owner's representative advise the contractor regarding the acceptability of any equipment that has been proposed for use on the job. This should be done *before the contractor brings the equipment on the site*. If it is left until later, rejection of equipment can result in additional expense to the contractor or to the owner.

Another particular point the foundation inspector should check is that the contractor's drilling tools conform to any requirements that are set forth in the specifications.

At the inspector's preconstruction meeting with the contractor, a list of persons who will receive copies of the inspector's reports should be made up, and copies of the list should be distributed promptly to all persons concerned.

4.4.4.   Coordination with Structural Engineer and Building Inspector

It would be helpful if the foundation inspector (and the geotechnical engineer if the two positions are separate) would meet with the designer of the structure and with a representative of the local building department before the work is started; however, this is almost never done unless a code deviation is requested.

The inspector should establish lines of communication (through proper channels) so that he can get prompt action if questions arise that he cannot resolve.

4.4.5.   Written Instructions to Foundation Inspector

Detailed written instructions, with a checklist of equipment to be taken to the job, should be furnished to the foundation inspector by the geotechnical engineer. If there is no geotechnical engineer assigned to the job, this is the responsibility of the resident engineer or the architect. It is important that these instructions should not direct or authorize the inspector to request the contractor to do more than is required or implied by the specifications and contract. The existence of clearly expressed written instructions, discussed and agreed to before inspection starts, is of prime importance not only in performing the job, but also in avoiding or resolving disputes or claims that might arise later regarding job performance.

4.4.6.   Inspection Equipment

The foundation inspector may need any or all of the following equipment, some of which will be needed for performance of his inspection duties and some for his comfort and safety:

Folding rule or measuring tape

Powerful flashlight

Hand mirror for reflecting sunlight down the hole

Sampling trowel, knife, jars or sacks, labels

Pocket penetrometer and/or torvane (or other instrument for measuring cohesive strength of soil)

Mud testing equipment for density, viscosity, sand content, and pH

Rock probe (Fig. 4.5)

Geologist's hammer for sampling and testing soundness of rock

Recording and report forms

Safety helmet

Raincoat or rain suit

Safety line and harness or cage

Miner's safety lamp or gas test meter

Air tank, fittings, and mask

Of course not all of these will be needed on every job, but they should all be available on any job site where they might be needed, and the foundation inspector should conscientiously check that all needed items are present and in good working order before he begins his work. Some of this equipment—particularly that concerned with safety—is usually provided by the contractor. Any such requirement should be stated in the specifications.

4.5.   Inspection of the Hole

4.5.1.   Safety Considerations—Entering the Hole

The first hazard encountered by the pier hole inspector is that of stones flung from the auger. With the open-helix type of drilling tool, soil and rock cuttings brought out of the hole are spun off the auger, to be picked up and removed by other equipment during the auger's next trip into the hole. Stones of considerable size are often brought out and flung outside the radius of deposit of lighter particles (see Fig. 4.2).

Surface inspection—either sampling and examination of spoil or looking into the hole—should be performed only when the machine operator has been made aware of the inspector's needs and intentions.

Entering the pier hole for inspection of cleanout can be very risky. It is imperative that the foundation inspector take all precautions that might be needed for his own safety, and he should advise the contractor and his employer promptly of any hazards or of any corrective measures that he considers necessary for his safety. He must never become careless, nor enter a hole when he has the slightest doubt of safety or when any desired precaution has been omitted or skimped.

Some of the hazards that may be present, and the safety precautions that should be taken, are discussed in the following paragraphs.

Caving or Collapse of Hole or Bell. This is the most obvious hazard, and one of the most dangerous. There are some geologic formations of such strength and uniformity that holes and underreams can be counted on to stand open without caving or sloughing. However, in the United States, OSHA regulations require that safety casing be used in all shafts that people enter, regardless of the supposed safety of the formation. In other countries, where local experience has led to customary inspection and cleanout without protective casing, anomalies are possible, and the inspector should be alert

FIGURE 4.2. Model SS-8487 crane attachment spinning spoil off a large auger. The men stand well back from the hole to avoid being hit by flung stones. (Photo courtesy of W-J Sales and Engineering Co., Houston, Texas.)

for danger and should insist on the use of protective casing in the shaft if he has any doubts at all or if the holes are of unusual depth or diameter. Slickensided or fissured, overconsolidated clays, or slaking shales, which drill cleanly and stand nicely for underreaming, are subject to dangerous movements of large blocks when exposed to wetting or drying, and particularly when they carry even very small amounts of water in pervious seams or layers, or in joints or fissures.

The protective casing should be positively supported at the top and should extend to within 3 or 4 ft of the bottom, or to the top of the bell, or to such other depth as the inspector may direct. For inspection of the sidewalls of friction pier shafts, windows can be cut in the protective casing; or the casing can be raised by stages.

Fall-in from Top of Hole. Before descending, the inspector should make sure there are no loose stones, clods, tools, or even piles of loose soil or rock that could fall or be knocked into the hole while he is in it; and he *must* wear a safety hat. If water is "raining" into the shaft or bell from water-bearing strata, he should wear suitable rain gear. In cold weather this is a safety measure as well as one intended for his comfort.

Potential Health Hazards. The authors know of at least one fatal accident due to natural or man-induced toxic substances encountered during pier drilling and have observed many incidents which could have led to fatalities. There are many known possible sources of substances which are dangerous to the health of field personnel via inhalation (most commonly) and direct contact: "bad air." Some of these that the authors have encountered are listed below:

1. Marsh gas from natural organic deposits.

2. Gas from organic material in fills (especially garbage fills).

3. Gas from volcanic sources.

4. Gas from natural gas or petroleum deposits.

5. Gas or fumes from industrial wastes or from leakage of volatile liquids from tanks, pipelines, and so on. (This source is not at all uncommon near gasoline stations and industrial plants.)

6. Accumulated CO_2 (sometimes mixed with CO) from motor traffic, or from any internal combustion motors, such as a battery of air compressors. A common source of CO and CO_2 in the hole is the use of a gasoline-driven bottom-hole pump. (These gases, being heavier than air, may also flow into a pier hole from the surface, displacing the air.)

7. Oxygen depletion—for example, from a previous descent, or from too long a stay in the hole (4.5).

Any of these conditions can be present without giving any warning by odor. Any of them will cause a man to lose consciousness, often without his being aware of it; and any of them will cause death in a few minutes unless the victim is removed to good air and, in some cases, is given artificial respiration. In every case where the possibility of bad air is suspected, fresh air should be introduced into the bottom of the hole by an air hose; and rescue means should be on hand for every descent into a pier hole, whether it is believed to be dangerous or not.

Hazard Detection and Risk Mitigation—Construction Safety. A hole can be tested for methane by using a combustible gas indicator (CGI) and for both combustible gas and oxygen by a dual purpose meter. Holes should not be entered if the methane concentration is 10% higher than the "lower explosion limit" or if the oxygen content is less than 19.5%. However, if toxic contaminants in the soil or groundwater are suspected, specialized testing is required. Under these circumstances the inspector must advise the owner or his representative to halt construction and retain a specialist (Industrial Hygienist) to assess the hazards and advise on appropriate safety procedures during construction. Should the level of contamination be found to be potentially dangerous to site personnel, and state and federal governmental regulations can be met, a health and safety plan should be prepared and implemented before construction is continued.

Safety measures that should be taken for every descent include the following:

Safety harness and line (Fig. 4.3) should be worn on every descent, even in shallow holes, so that an unconscious man can be hoisted out of the hole without delay and without the danger of injury inherent in the use of improvised rope slings. The practice of sending a man down to lift and carry out an unconscious man, or to hold him while the two of them are being hoisted out, is dangerous and unsatisfactory.

Smoking in the hole is unsafe and should never be done.

In general, self-contained breathing apparatus (SCBA) should be part of the foundation inspector's kit and should be strapped on and arranged for quick application (not hand-carried) on every descent (see Fig. 4.4).

FIGURE 4.3. Wrist-type rescue harness (Mine Safety Apparatus Company, Pittsburgh, Pennsylvania).

FIGURE 4.4. Hip-air breathing apparatus to be used with air hose (Mine Safety Apparatus Company, Pittsburgh, Pennsylvania).

An observer at the top of the hole should be present at all times when a person is in the hole. It is not sufficient to assign this duty to the drilling machine operator; he cannot properly operate his hoist during a rescue operation and at the same time be sure that the victim is lifted freely and without injury. There should be reliable means of communication between the man in the hole and the observer. In very deep holes a wired or wireless telephone system should be used.

An electric light of the safety lamp type, with the cord in good condition, should be suspended in the hole in such a position that the inspector can move it to inspect any part of the hole.

For safety during descent into holes 3 ft or more in diameter and not more than 20 ft deep, a ladder may be the most convenient means; but in most instances some sort of hoisting equipment should be on hand for rescue purposes in case of trouble. It is almost impossible to carry an unconscious man up a ladder in a small-diameter hole.

It is fairly common practice for the drill operator to let a man stand on a rod or pin through a hole in the bottom of the kelly bar and hold onto the bar while it is lowered into the hole, or to sit or stand in a sling or bosun's chair raised and lowered by the drum line on the rig. Unless there is a "fail-safe" arrangement on the kelly hoist or drum line, either of these is a risky procedure. Cable- or chain-operated kellys should never be ridden by personnel, nor should cables on mechanical winches. A positive forward and reverse hydraulic winch is relatively safe; so is a power-up and power-down hoist line on a crane. A hand-operated windlass with a ratchet to prevent accidental release, on a tripod set up over the hole, is usually safe.

The Contractor's Responsibility. Whatever means is used for entering and leaving the hole, the contract specifications should state that it is the contractor's responsibility to furnish it; and the conveyance should be convenient, safe, and not uncomfortable for the user. It should also be the contractor's responsibility (and this should also be stated in the contract) to take all applicable safety precautions, including any requested by the foundation inspector, and to furnish any equipment or supplies needed for such precautions (beyond those in the inspector's kit). The need for such equipment is one of the items the inspector should discuss with the contractor before the job is started. Anything that might be required should be on the job site and readily available when needed, and no one should have to hunt for it.

4.5.2. Identification of Bearing Stratum

The plans should indicate the stratum or strata of soil or rock in which it is expected that the pier holes will bottom. Usually, the descriptions will be taken from the soil and foundation report and will be worded so that identification of the various strata as they are penetrated by the auger should be easy. Occasionally, plans will show boring logs made out by someone without geologic or soil identification training, and soil and rock descriptions may be uncertain, or confusing, or even incorrect and misleading. (A common example is the designation of an inorganic silt as a "clay"—a dangerous error because silt and clay behave quite differently under load and during excavation.)

As mentioned earlier, the foundation inspector should examine the soil samples or rock cores in advance, become familiar with the appearance and "feel" of the proposed bearing stratum, and take typical core and soil samples, the latter *sealed in jars at natural moisture content*, to the job site for comparison.

Occasionally, when pier holes are very shallow and when the bottom is clean and there is no water in the hole, it is sufficient to inspect a hole from the surface, examining the sides and bottom by reflected sunlight from a hand mirror or, on dark days, with a powerful flashlight. The condition of the proposed bearing surface can often be judged by probing the bottom with a rod. Identification of the formations penetrated and of the bearing stratum can usually be made from the cuttings as the drilling proceeds. Examination of the larger intact lumps of soil or rock brought out by the auger will give some information about the *in-situ* firmness or strength of the formations. Inspection from the ground surface is not generally recommended, but is the only means available in holes smaller than 24 in. in diameter. (Thirty inches is the practical minimum.)

In the normal pier-inspection process, the foundation inspector will have to descend into the holes and examine the proposed bearing surface.

The foundation inspector must always be on the alert for variations in stratigraphy or for soil or rock conditions that are not in agreement with the reported boring information. Most test borings are not sampled continuously, and continuity of strata between borings is never certain. It is not unusual for significant variations in subsurface conditions to be missed in the test borings, but show up in the pier holes and require on-the-spot decisions by the foundation inspector.

The function of identifying the bearing stratum is one of the most important and critical tasks the foundation inspector has to perform. Obviously, the safety of the structure depends on the foundation's resting on suitable support. More than safety, however, depends on the foundation inspector's performance of this task. Drilled pier jobs are priced on the basis of an estimated average depth and a range of depths for the pier holes. Every foot of depth past those shown on the plans will increase the cost of the foundations, often beyond the owner's budget. The foundation inspector must not hesitate to advise the contractor if the right material (or condition) has not been reached, but he must also be positive in identifying suitable support when it has been reached. Bearing elevations are determined by on-site inspection, and rarely conform to the elevations *estimated* for design. This should be carefully stated in the geotechnical report by wording such as: "bearing elevations listed herein are for the purpose of providing estimates for bidding purposes; actual bearing elevations will be determined in the field at the time the pier holes are drilled." Whenever the possibility of large overruns in pier depth becomes evident to the foundation inspector, the owner or his representative should be notified at once. Many present-day pier-drilling machines are very powerful, and will cut very competent formations with such ease that there is danger of drilling on past the designated bearing stratum before it is recognized, unless the anticipated change is from soil to hard rock. Some machines will even drill relatively hard rock with comparative ease.

4.5.3. Inspection of Water- or Mud-Filled Holes

It is always preferable to inspect, and to place concrete, in a dry hole; but instances will be encountered where dewatering the hole proves impractical.

For piers in soil, this contingency can be handled by drilling in a water- or slurry-filled hole (Chapter 3). In a water- or mud-filled hole, inspection is limited to observation of drilling, examination of cuttings, and probing the bottom with a rod. This is an incomplete inspection; cuttings will be softened and may be so altered or contaminated as to reduce stratum identification to guesswork. The risks from inspection under these conditions can be minimized by care. They cannot, however, be eliminated.

Wherever concrete is to be placed in a slurry-filled hole, hole geometry and volume must be determined as accurately as possible. The viscosity and sand content of the drilling slurry *must be controlled and verified by the inspector*, so that complete displacement of slurry by concrete is assured; and the volume of concrete placed must be determined very accurately.

In pier holes in rock where dewatering proves impracticable and it is planned to use pumped-in or tremied concrete, inspection can be much more thorough. If the completed hole can be flushed out with clear water without affecting the hole stability, sidewalls and bottom can be examined in detail by means of a closed-circuit television camera. This means of examination is especially applicable to inspection of holes of great depth.

4.5.4. Bottom Cleanout or Stabilization

Any drilled pier hole should be inspected to assure that there will be good bearing contact between the concrete and the proposed load-bearing surface. Many holes require hand cleaning of the bottom, and holes for piers that depend on sidewall shear for part of their support should be inspected for clay smear on the walls, or any other condition that could interfere with the development of that support. Where grooves in rock walls are required, the foundation inspection must make sure that they are properly cut and are clean. Inspection of bottom should immediately precede concreting, because sloughing of bell or shaft wall, fall-in from the surface, or water-softening of an initially sound bottom surface could possibly alter the bearing condition enough to lead to damaging postconstruction settlement.

Ideally, the bottom of a pier hole should be clean. The geotechnical engineer must be the judge of how much loose material—or mud or "slop"—can be left on the bottom without risking later settlement of the loaded pier as the soft material in the contact zone consolidates under load. The foundation inspector's written instructions should give him direction on this point, and he should not hesitate to consult the geotechnical engineer or supervising engineer if he is in doubt about the acceptability of the bearing surface. (See also the discussion on pp. xx–xx.)

When ground conditions are such that the contractor finds it impossible or very difficult to obtain and keep a clean bottom until the concrete is placed, the inspector should immediately consult the geotechnical engineer about a change in procedure or design.

The foundation inspector should report any tendency to leave a hole open longer than necessary, especially in underreamed holes. An experienced contractor is not likely to do this, because it means trouble in cleanout, and under some circumstances can mean complete collapse of the hole or bell, and redrilling to a deeper bearing level at the contractor's expense. Incidents of this nature should be described in the foundation inspector's daily reports, and be reported promptly.

4.5.5. Proof-Testing

Rock Bearing. The need for proof-testing of rock-bearing surface depends upon the possibility of less competent zones of rock, or voids, occurring within the depth of effective stress increase below the readily inspectable bearing surface. The need for proof-testing should be determined in advance, if possible, based on knowledge of local subsurface conditions and information obtained in the foundation investigation and presented in the soil and foundation report. If proof-testing is considered necessary, the method and frequency of testing should be detailed in the contract documents, and there should be a provision in the contract for increase or decrease in the estimated work involved. In the authors' practice, proof-testing is usually required for all pier holes in rock formations subject to significant and irregular weathering, where the unit loads to be imposed on the bearing surface are 10–15 tsf (N/m^2) or more. Some solution-prone limestone formations are so seamed, fissured, and cavernous that every pier hole may have to be proof-tested regardless of the design loading.

Proof-testing of rock usually consists of drilling an exploratory hole, about 2 in in diameter, in the bottom of a pier hole that has reached a supposed bearing level, and observing such indicators as speed of drilling under a given drill pressure, dropping or clogging of bit, loss of drilling water (if used), and continuity of the bearing rock as judged by probing (or scraping) the sides of the completed probe hole with a right-angled chisel point formed at the end of a ⅜- or ½-in steel rod (see Fig. 4.5). Sometimes the proof-test drilling is done with a rotary diamond-coring machine, but more commonly a percussion rock drill with a star bit is used. With the latter machine, the foundation inspector should see that there is near-constant weight on the bit, and that a sharp clean bit is used, so that the speed of penetration of the drill can be compared between test locations. Successively smaller bits should be used if necessary to prevent drag of the drill steel (rod) or clogging as the test hole is deepened.

The recovery of rock core is preferred, but is significantly more expensive.

This method of percussion proof-testing is crude, but fairly effective. In such formations as limestones and dolomites, and the weathered mica schists and gneisses of the eastern Piedmont region, the authors consider it adequate and necessary for heavy loads or important structures. In the shales of the midwest, this test is not used. In these rocks, caverns and solution channels are very rare or absent, and the rate of penetration of either type of drill is not a good index of hardness because it is governed more by the stickiness of the clayey cuttings than by the hardness of the formation. In these formations, identification of the formation and the depth of "refusal" to the standard penetration test are, under some circumstances (e.g. compaction shales and weathered sedimentary rocks), considered adequate in lieu of any deeper proof-testing.

Chisel point

FIGURE 4.5. Rock probe rod made from ½-in. reinforcing bar.

The foundation inspector should record the type of drilling machine used and the type and condition of the bit. He should explain the desired proof-testing procedure to the driller, record any relevant observations that the driller can pass on to him, and record the time for each 6 in. of penetration. He should decide, on the basis of the criteria established by the geotechnical engineer, when to stop the drilling and make his sidewall probing test. If the probing shows lack of continuity of competent rock, the foundation inspector will state that satisfactory bearing has not been reached and recommend deepening the pier hole.

Cost of Proof-Testing. The present (1984) cost of proof-testing in rock by drilling techniques has been found to be of the order of $20 to $50 per test. Because the unit loads that could be approved if the proof-testing were not used would be only a relatively small fraction of the design loads that can be confirmed by the tests, it is obvious that proof-testing (where needed and applicable) is not an extra cost but an economy measure.

Penetrometer or torvane tests in clayey soils should not be considered an extra operation at all. The tests do not require anything of the contractor or delay his operations, and they do not take any more of an inspector's time than the "thumb" test he would make if he did not have the field instruments.

Soil Bearing. An experienced foundation inspector who has familiarized himself with the appearance and "feel" of the test boring samples from the proposed bearing stratum should be able to detect a weaker phase of this stratum from his examination of the auger cuttings and of the bottom of the hole. However, the use of a pocket penetrometer, a "torvane," or some other field instrument for providing an index to the in-situ strength of clayey soils is recommended. The consistent use of such a tool helps keep the inspector's judgment of clay strength from "drifting" and provides a valuable record to back up the inspector's notice to the contractor that bearing level has (or has not) been reached. A word of caution must be introduced here, however. Handheld instruments for measuring in-situ strength of clays may not give proper indications of the shearing strength of some clays; they are not at all appropriate for cohesionless soils (silts or sands). Formations containing cohesionless soils at or below bearing level require careful evaluation by a geotechnical engineer. Proof-testing techniques which have proven applicable are penetration-resistance tests using SPT or CPT methods as described in ASTM D 1586 or D 3441, respectively, or the portable plug sampler which employs a 30-lb hammer manually dropped 18 in. Note that interpretation of penetration-resistance tests must consider the effect of confining stresses which will be restored upon completion of the pier construction.

4.5.6. Proof by Load Test

On the occasional job where plate load tests or full-scale load tests are required to confirm design criteria, or to justify design changes involving changed criteria, the tests should be carried to plunging failure if at all possible. The number and method of tests should be specified in the contract documents or, if necessary, be covered by a supplementary agreement. The setup and actual performance of these tests are the responsibility of the contractor. The geotechnical engineer and the foundation inspector should serve as observers on these tests, recording the test data, deciding when load increments should be added or removed, and plotting and analyzing the test data. A practical and relatively inexpensive full-load test for completed piers has been described by J. O. Osterberg (patent applied for), and is covered in more detail in Subsection 3.13.2.

4.5.7. Determination of Location, Plumbness, and Dimensions

Alignment and plumbness of the pier holes are the responsibility of the contractor. To check the agreement of these measurements, as built, with the plans and specifications may be one of the assigned duties of the geotechnical engineer, and these duties are often delegated to the foundation inspector. Plumbness is easily checked, using a plumb bob suspended from the top of the hole, but alignment—that is, position of the center of the shaft, at the top, with respect to the design position—often is impossible for a foundation inspector to determine accurately. The geotechnical engineer should not accept this responsibility unless the owner agrees to furnish batter boards (or other suitable location markers) that are not disturbed by the pier-drilling and concreting operations. Such markers should be so positioned as to allow the needed measurements to be made on the seated casing and on the completed piers. It is the contractor's responsibility to see that the batter boards or other location markers are not disturbed.

When an uncased pier has been concreted, it may be impossible for the inspector to determine the exact location of its center, for such pier holes are rarely completed without some irregular raveling at the surface. The remedy is to set a short piece of surface casing or liner, to be left permanently in place, before the pier is concreted.

Insofar as is practicable, the foundation inspector should inform all parties as soon as it appears that a drilled pier hole does not conform to the specified tolerances, so that corrections can be made before concrete is placed. He should notify the owner's representative (through the geotechnical engineer) immediately whenever it is determined that a completed pier is not correctly located, so that design of corrective piers or other measures can be started. Any avoidable delay in getting such corrections started might be used by the contractor as the basis for a claim for an "extra."

Occasionally ground conditions will be such that the contractor will be consistently unable to complete piers within plumbness or location tolerances. As soon as this condition is noted, it should be reviewed with the geotechnical engineer and the owner's representative, so that the possibility of relaxing the tolerances, or of redesigning the piers to permit larger tolerances, can be considered. A few inches increase in shaft diameter, for example, might be less expensive and more acceptable to all concerned than the use of additional piers to compensate for misalignments.

4.5.8. *Guarding Against Lost Ground and Subsidence*

A drilled pier hole in ideal ground conditions does not produce appreciable movement of adjoining ground. There is a tendency to think of all drilled pier installations as being in ideal ground conditions, probably because contractors are increasingly able to cope with unfavorable ground conditions. The potential for lost ground is often overlooked until subsidence and/or surface caving adjacent to a pier has occurred, sometimes with serious (though usually localized) damage to nearby buildings, streets, or utilities. Subsidence caused by the effects of groundwater drawdown can affect relatively large areas, particularly where the subsoils are highly compressible.

The foundation inspector should be alert for ground conditions that could produce inflow of excess soil as the drilling is done, and for detrimental effects of groundwater drawdown. He may also be instructed to examine nearby surfaces and structures before pier drilling is started. If he has this duty, existing cracks in buildings, ground, or pavement should be documented by dated and signed photographs, sketches, and notes, which should be part of the foundation inspector's first reports. For major works, this documentation requires professional photography and surveying.

Inspection for evidence of lost ground should be repeated daily. Initial indications of ground loss potential, as well as of unobserved hole caving, can be obtained from comparison of the volume of concrete poured in a hole with the theoretical volume required to fill the hole. Indications of lost ground may take the form of subsidence immediately adjacent to a pier location, new or widening cracks, curb separating from pavement, soil pulling away from nearby foundation walls, or development of sags or "bird baths" in adjacent pavement or lawn. An excess volume of soil may not be evident as it is removed from the hole; but there may be indications during the drilling noticeable to the drilling-machine operator, such as closing in of the hole, sand inflow, or soil in the water discharged from dewatering pumps, whether they are pumping from a pier hole being dewatered for concreting, from a deep-well drawdown system, or from a well-point installation.

The foundation inspector should also observe any groundwater variations that occur during construction and be alert to the possibility of ground movement due to such variations.

At the first indication of potentially damaging ground loss, the foundation inspector should make measurements and notify the geotechnical engineer, so that recommendations can be made for changes in construction procedures or other special measures can be taken to prevent damage to existing structures.

4.6. Reinforcing Bar Cage Inspection

Inspection of reinforcing bar cages for size and condition of bars and dimensions of cage is not always included in the duties of the foundation inspector. However, the correct *positioning* of the cage in the pier is an important matter, and the foundation inspector is on hand when the cage is placed in position and while the concrete is being placed. The contract documents should state clearly who is responsible for observing and reporting on the positioning of the cage, the means by which it is held in position, and its apparent position after the concrete has been placed.

Where rebar inspection is required, care must be taken to verify that the bar sizes, weights, and connection details conform to the plans and specifications. The inspector must also verify that the rebars are free of rust or other contaminants that could reduce bond strength. During placement of concrete and casing withdrawal (if relevant), the vertical and torsional movement of the cage, as well as its initial positioning, must be monitored carefully.

4.7. Inspection of Concrete Placement

4.7.1. General Objectives

The general objectives of concrete inspection on a drilled pier job are (1) to see that the concrete has the specified strength and (2) to make sure that the concrete of the pier is continuous, in full design section, from the bottom to the top of the pier. The first of these objectives is common to all concrete jobs. The task of taking test cylinders and making slump tests may be assigned to a foundation inspector, but is more commonly contracted to a commercial testing laboratory that has the responsibility for quality assurance of the concrete. The requirement of assuring integrity and continuity of the finished pier is extremely important. It requires specialized experience and knowledge and should be handled by someone qualified as a foundation inspector. When a drilled pier is completed, whatever has been done is buried. There is no opportunity later to strip forms and see how it looks or to use a "soundness hammer" on the sides of the column to check the integrity of interior concrete. Although observation and reporting of the placement of concrete may be an assigned duty of the concrete inspector, this function is so important that it should also be a responsibility of the foundation inspector.

4.7.2. Water in the Hole

If possible, the pier hole should be clean and dry when concrete is placed. In many instances, however, complete absence of water in the hole will be impossible. This contingency should be anticipated in the design and covered in the specifications. The allowable depth of water will depend on (1) the type of pier, and (2) the method of concrete placement. Two to three in. (5–8 cm) of water in the bottom of a straight pier hole can usually be displaced or mixed with free-falling concrete without causing serious segregation or weakening of the concrete; but the same amount in a belled hole will have much greater effect, and may, as it is displaced and rises into the shaft, cause complete loss of strength in the concrete at the junction of bell and shaft. The use of clean coarse gravel, or of dry cement, to "blot up" a limited amount of free water is sometimes permitted, but either device should be used with caution.

When concrete is placed by tremie or by pumping, with the hose at the bottom of the hole, free water in the hole has little effect.

The foundation inspector must know what water depths the specifications will allow, as well as the method of placement specified. He must be prepared to insist on (1) effective dewatering measures, or (2) tremie placement or pumped-in concrete, in case excessive water enters the hole. Approval must be obtained from the geotechnical engineer if deviation from the specifications is involved, and the approved deviations must be recorded in the foundation inspector's written reports.

Before concrete can be placed under water, either by tremie or by pumping, the water level in the hole must have reached a stable level; otherwise, water will continue to flow into the hole, washing or diluting the concrete, until the concrete level has reached a point where its pressure exceeds the outside water pressure at the point of entry. However, recent research (4.6) has shown that the hydrostatic pressure exerted by fresh concrete is not necessarily the same as the hydrostatic pressure calculated from an assumed weight of, say, 145 pcf. In all of the tests, the lateral pressures exerted by the fresh concrete depended on the slump, on the rate of placement, and on the elevation of the concrete surface above the base of the shaft. These observations suggest another reason for requiring high-slump concrete whenever outside water pressures are troublesome.

The specifications should specify a minimum size of tremie pipe, and the inspector should see that this limit is observed. If this item is not covered in the specifications, he should not approve the use of a tremie pipe smaller than 8 in. (20 cm) in diameter, and 12 in. (30 cm) is preferable. (Note that this limitation applies to the pipe for *tremied* concrete, but not to that for *pumped* concrete.) The inspector should keep in mind that tremied concrete will add $12 to $14 per cubic yard to the cost of the pier, and that a good casing seal and dry hole are a better solution where they can be obtained economically.

4.7.3.   The Concrete

The specifications will state allowable limits for the slump of the concrete to be used. This means, of course, the slump at the time the concrete is placed. If there are delays in delivery, or in placing after the concrete truck reaches the job, the concrete may have set up enough that its slump is outside the lower limit; and in such case the foundation inspector *must* reject it. There may be a separate concrete inspector on the job, who will make the slump tests; but it is the foundation inspector's duty to reject the batch if the slump is too low. To accomplish this, to plan the deliveries and placement so that the hole can be concreted during the time that the slump is adequate, the foundation inspector needs to have, from the concrete supplier, a slump versus time curve for the mix design used; and he needs to check the actual times of placement against this curve as the concreting proceeds.

4.7.4.   Slurry Testing and Control

Whenever the concrete is to be placed in a slurry-filled hole, control of the viscosity and sand content of the slurry at the time of placement (by tremie or by pumping) is necessary, to avoid the possibilities of contamination of the concrete, or of incomplete displacement of the slurry (see Subsection 3.5.7). The foundation inspector should be responsible for testing and control of the slurry at this very important stage of the operation.

4.7.5.   Final Inspection Before Concrete Placement

Although a pier hole may have been inspected and found clean and dry and ready to be filled with concrete, if the hole is potentially unstable and if more than a few minutes have passed after inspection, it should be inspected again immediately before the concrete is placed. Unexpected sloughing or caving of a bell or uncased shaft can also occur in what appears to be a stable shaft, so any significant delay should be followed by a reinspection. There have been serious instances of deep collapse of permanent casing produced by soil and water pressure, undetected until the pier had been completed and partly loaded (3.29). Immediately before concrete is placed, the foundation inspector should take a final look at walls, bottom, and reinforcing cage, using a powerful flashlight.

4.7.6.   Cold-Weather Concreting—Fog in the Hole

Often when a deep-drilled pier is to be completed in cold weather, fog will form in the hole and interfere with the final inspection from the top. This may occur before concrete placement is started. It is likely to happen as soon as the relatively warm concrete begins to enter the hole. In either case, the presence of fog must not be accepted as an excuse for omitting or skimping on the final inspection of the hole and the observation of concrete level as the hole is filled or as casing is pulled. If necessary to allow inspection, the fog must be dispersed by lowering a heater or blowing warm air down the hole. If cold-weather concreting of deep piers is anticipated, this item should be provided for in the specifications; otherwise the contractor may be unprepared and be entitled to an "extra."

4.7.7. Avoidance of Segregation

To avoid detrimental segregation in concrete, the specifications usually provide that the concrete shall be discharged through a hopper having a bottom spout centered over the hole to concentrate the falling concrete in a stream of small diameter compared to the hole or reinforcing cage diameter, and that the hopper shall never be permitted to empty until the hole is filled with concrete. The foundation inspector must see that this provision is followed, or that a satisfactory alternate method of avoiding segregation, such as the use of an "elephant's trunk," a tremie pipe, or pumped-in concrete, is employed. If an alternate method is used, it must be approved in writing by the owner's representative (or the geotechnical engineer) and must be documented in the foundation inspector's reports.

4.7.8. Concreting in Reinforcing Cages

Concrete slump and maximum size of aggregate must be carefully designed to assure that either free-falling or pumped-in concrete will flow freely between the reinforcing bars and completely fill the space outside the cage. Although these are details normally covered by the structural engineer and his concrete inspection agency (usually a commercial testing laboratory), they are so vital to the performance of the pier that the foundation inspector too should be alert to report the placement of any concrete that does not flow freely through the openings in the cage. The foundation inspector will judge this by the appearance and behavior of the concrete and by the volume of concrete required to fill the hole. Concrete with less than 4-in. (10 cm) slump will not flow readily between closely spaced reinforcing bars, and a properly designed mix with a 6- to 7-in. (15- to 18-cm) slump is preferable, particularly where heavy reinforcement is used. The inspectors should never accept concrete that has had water added to it to increase its workability.

Under some circumstances, reinforcing cages can become displaced or distorted during concrete placement or during the pulling of temporary casing (see Subsection 3.12.9). This fault may appear as a rise of the cage above its design position, or the cage may "squat" and even disappear below the surface of the concrete (Fig. 3.16).

A foundation inspector must be alert for the conditions that lead to this type of problem, and should if possible watch the top of the cage during removal of the casing as well as watch the concrete surface in the casing. When long lengths of casing are used, direct observation becomes impossible as the casing is hoisted; and the foundation inspector will have to form his judgment from his observation of the top of cage and concrete before and after removal of each casing length, the smoothness of the casing pulling operation, and the fluidity of the concrete. If a reinforcing cage disappears or is obviously distorted or displaced, the geotechnical engineer and the owner must be notified at once and instructions for correction be transmitted from the owner to the contractor.

4.7.9.   Vibration

The extent to which vibration is to be used may be covered by the specifications. If it is specified or used, it becomes one of the items that both the concrete inspector and the foundation inspector must observe and report.

Vibrators generally are either electric or air-operated and models of both are available for use in deep holes. The foundation inspector should include in his report a statement of the kind and model used, its operating condition, and how it was used—at what depths, length of time in operation, and so on. It is not sufficient merely to note "vibrated" unless a prior report gives the pertinent details.

4.7.10.   Interrupted Pours

Good practice requires, and the specifications should require, that a pier hole be filled in one continuous pour. The specifications should provide for a special procedure to be followed when this cannot be done. Occasionally, due to causes beyond the contractor's control, concrete placement will be interrupted. Procedures for assuring continuity of the pier concrete in such cases are described in Subsection 3.12.10 and the foundation inspector should be familiar with these procedures, in addition to the specific wording of the specifications. The foundation inspector should observe and record these operations and the time at which they occur, and he should refuse to approve any pier in which they are not performed as specified.

4.7.11.   Pulling Temporary Casing

This is an operation that is most critical, and one in which probably more mistakes are made than in any other in drilled pier construction. The foundation inspector must be present, alert, and aware of the potential for trouble inherent in the operation and in the particular formations that have been cased off.

When holes have been cased through formations that normally stand open, temporary or protective casing may be pulled after the bottom has been cleaned and inspected but before the concrete is placed. The authors recommend limiting this practice to relatively shallow holes in stiff clay formations. In this event, a clean hole should be exposed and the concrete should be placed immediately, and the inspector has only to watch for sloughing before or during the concrete placement. But in most instances, casing will not be pulled until it has been partially filled with concrete; and to assure the continuity of the completed column of concrete, the inspector has to watch for any occurrence that threatens the integrity of the completed column.

Especial care is needed when there is a groundwater level outside the casing, which must be held back by the fluid pressure of the concrete inside the casing when the bottom seal is broken. Construction errors that can be made under these circumstances are discussed in Subsection 3.12.9.

The foundation inspector should make every effort to know where groundwater level is outside the casing, and if there was any inflow from a pervious stratum (or rock fissure or cavity) during drilling, he should be aware of it and should report it.

The casing *must* be filled with concrete to a level at least high enough to balance the groundwater level outside the casing before it is lifted off its bottom seal. (The inspector should keep in mind that 2.4 ft (73 cm) of water will be balanced by 1 ft (30 cm) of concrete.) This will be sufficient to keep a positive head of concrete against inflow of water at the bottom of the casing as it is being raised. The relation between required concrete pressure and existing groundwater pressure is illustrated in Fig. 3.19.

A word of caution: The specifications should require that temporary casing—especially telescoping casing—be clean and free of dents, or else there will be a tendency to hang up and bind during removal. Some specifications require also that the casing should be well oiled. (Some contractors insist that oiling is a waste of time, as all oil is wiped from the metal surface by the time the casing is inserted and concrete placed.)

The foundation inspector should call the contractor's attention to any deviations from these specification requirements, and should indicate such notice, with a record of compliance or noncompliance, in his daily reports.

Vibratory pile pullers have recently been adapted to pulling drilled pier casings and have been found very effective. With vibration, casing can be lifted smoothly at a controlled rate, and the vibration probably helps develop good sidewall shear contact and good bond between concrete and reinforcing steel. An added dividend—as far as the contractor is concerned— is that the use of the vibrating puller reduces the danger of overstraining crane booms, and thus increases job safety.

4.8.   Obligations of Other Persons

For a smooth-running drilled pier job, there are obligations and qualifications which appertain to other persons as well as to the geotechnical engineer and the foundation inspector. As the authors see these, they are as follows.

4.8.1.   Owner

The owner has an obligation to employ competent engineers and inspectors, not on the basis of cost but rather on the basis of qualifications. He should define their obligations and duties clearly, and back up their decisions and recommendations promptly.

4.8.2.   Contractor

The contractor's prime responsibility is to complete the contract on time, in accordance with the plans and specifications, and in a workmanlike manner.

One of his obligations, which should be explicitly set forth in the contract, is to correct promptly all specification deviations reported by the owner's representative. Another is to do his work in such a way as to minimize engineering construction review and inspection costs. Experience with contractors readily separates those who can be expected to turn out a well-managed job from those who can be expected to end the operation in avoidable disputes, arguments, and claims for "extras" or "changed conditions." Some of the qualifications that the geotechnical engineer looks for in a drilled pier contractor are adequate work force, suitable equipment, freedom from overcommitment on other contracts, experience with drilled pier construction under conditions similar to those at hand, willingness to make corrections and to change unsatisfactory construction practices when they are called to his attention, and a record for finishing jobs on time and with a minimum of disputes or claims for "extras." (Of course, disputes and claims for "extras" are not necessarily the contractor's fault. Sometimes the grounds for disputes and claims are written into the contract; sometimes they are incorporated in the plans and specifications; sometimes they show up as unexpected subsurface conditions; and sometimes—unfortunately—they are the result of bad decisions by engineers or inspectors.)

4.8.3. Building Department Inspector

This official has the obligation of seeing that any construction in his jurisdiction meets the requirements of his building code and, often, the requirements of departmental rulings or practices that may comprise interpretations of the building code. His duty is to protect the public's safety. One of his obligations is to make his inspections promptly when notified. In order for him to do this, of course, someone must keep him informed of the progress of the work and give him adequate notice of occasions when the work will be ready for his inspection. For drilled pier construction, the foundation inspector should keep this in mind. A good foundation inspector will earn the confidence of the building inspector.

4.9. Control of the Job

As has been indicated many times in the preceding pages, the construction of drilled pier foundations requires suitable design, careful construction, and good and knowledgeable control of construction details if both the supporting capacity and the potential for economy of this type of foundation are to be realized. This is true of any foundation construction, of course, but both the supporting capacity and the cost of drilled pier construction are more sensitive to variations in these aspects than are most other types of foundation. Mistakes in design or in construction details can easily be overlooked until too late, resulting in unnecessary additional costs and construction delays. On projects involving drilled piers, both design and construction decisions (and changes) may be required as the work proceeds, and therefore close job control is imperative. This means, of course, that the geotechnical engineer and the foundation inspector must be competent. It is equally important that the contractor must be cooperative (as well as capable). Finally, the owner should realize the importance—to him—of what the construction review team is trying to accomplish, and should be prepared to take immediate steps to back up their decisions if the necessity should arise.

To implement close control, the inspector has to report promptly and regularly. The lines of communication between inspector, geotechnical engineer, structural engineer or architect, and owner have to be established and maintained; and both the geotechnical engineer and the inspector have to earn and keep the confidence of the owner and his agents and the respect of the contractor.

A message to owners: competent engineering construction review and inspection do not increase job costs. They save both money and time.

4.10. Postconstruction Inspection

There have been some large-scale construction errors in recent years which resulted in the necessity for replacing some very large and expensive piers in which the completed concrete shaft was found to be discontinuous or locally reduced in cross section. As a result, it has become customary in some areas to check the integrity of a representative sampling of completed piers on important jobs by making diamond core borings for the entire length of the shaft; or, in some cases, by making a boring outside but parallel to the shaft in order to confirm the presence or condition of the bell (3.17). This test is applied also to piers whose integrity is suspect because of construction circumstances or inspectors' observations (see also Subsection 3.6).

References

4.1   Loss Prevention Manual, *Consulting Eng. Council U.S.A.*, 1969.

4.2   Davidson, David McL., The Legal Implications of Quality Control, *Civil Eng.*, November 1967.

4.3   Goldbloom, Joseph, Safety in Construction—Whose Responsibility? *Civil Eng.*, November 1969, p. 42.

4.4   McKaig, Thomas H., *Field Inspection of Building Construction*, McGraw-Hill Book Company, New York, 1958.

4.5   Poisonous Subsoil Air, *ENR*, August 26, 1971, p. 13.

4.6   Bernal, Juan B., and Reese, L. C. Drilled Shafts and Lateral Pressure of Fresh Concrete: New Research Findings, *Foundation Drilling*, May 1984, p. 10.

5

Exploration for Drilled Pier Foundations

5.1. General Requirements

The authors' experience has led to the conclusion that more careful and thorough foundation exploration is required for drilled pier foundations than for most other deep-foundation systems. Part of this conclusion derives from the sensitivity of drilled pier costs to apparently minor geologic variations, and part from the difficulty of anticipating pier construction contingencies from the results of conventional test borings. This conclusion is probably not correct for areas where the bearing formations for drilled piers are uniform and readily identifiable rocks or soils, and where drilling conditions are generally favorable.

It is imperative that the exploration program include enough borings, soundings, or test pits to establish a strong inference regarding the continuity (or lack of continuity!) of the formations penetrated throughout the area in which the pier holes will be drilled. The test borings must extend into the proposed bearing stratum a distance sufficient to establish the adequacy of that stratum within the depth of significant stress increase imposed by the pier.

The person supervising the exploration program needs both knowledge of the proposed structure and understanding of the behavior under load— as well as during drilling—of the soil and rock formations penetrated. For the benefit of the foundation contractor, good representative samples must be obtained and preserved at natural water content, for inspection later by prospective bidders. "Undisturbed" samples may have to be taken for strength and consolidation testing. The borings should be logged accurately and completely, in such terms that the prospective bidder can recognize the materials penetrated and anticipate their behavior under his equipment.

Because of the important influence of apparently minor variations in materials, continuous sampling may be necessary in selected borings. "Undisturbed" as well as frequent drive sampling is appropriate within cohesive soil strata that might be expected to stand open without casing, and especially within strata in which underreams may be considered.

5.2. Site Reconnaissance

An important part of the geotechnical report should be a physical description of the site and its vicinity. This should be based on a ground reconnaissance by the geotechnical engineer, and should cover the following questions and observations:

Access problems for equipment—trucks, cranes if required; other equipment; gates, roads.

Drainage—Will poor trafficability develop in bad weather? Are there ponds or streams?

Nature of surface—Is there landfill? What kind? How deep?

Utilities to be moved or protected from the work and from construction traffic?

Vegetation—trees? How large? Brush?

Rock exposures—not only on the site, but in the vicinity? Identify.

Structures to be removed? If so, with basements? Deep foundations (if known)?

Nearby structures that might be affected by the proposed work?

Research into Past Site Development is also crucial in assessing possible subsurface obstructions, and can provide useful information for the reconnaissance.

5.3. Conventional Exploration

Test borings—Any type of test boring that is suitable for retrieval of uncontaminated soil and rock samples can be used; but especial care should be taken in the logging, the preservation of samples, and the accurate determination of groundwater levels. Because of the latter requirement, auger borings through the overburden, using hollow-stem flights where unstable soil conditions are anticipated, are especially suitable. Cased chop-and-wash and rotary borings, using either water or drilling mud to carry out the cuttings, are likely to have a tendency to seal the sidewalls of the test hole and delay correct indication of true groundwater level, or may obscure detection of water-bearing strata.

Soundings can be used to augment the test borings in establishing the depth to rock or hardpan, the thickness of hardpan or of a weathered rock zone over bedrock, and sometimes the condition of the rock. For this purpose, soundings can be unsampled auger, wash, rotary, or air-drilled holes which may be carried to or into rock. Soundings have also been made using probes to rock using a hammer-driven or hydraulically pushed rod, often equipped with an expendable cone point larger than the diameter of the rod.

Pneumatic percussion or rotary-percussion drill soundings have particularly good utility in investigating rock quality by recording the rate of penetration under a near-constant drill pressure. By using a digital data logger or strip chart to record the drill-rod travel in real time, an accurate rate-of-penetration log can be obtained which provides excellent correlation with stratigraphic changes and rock discontinuities. Instrumentation of the bit pressure and drilling torque have also been shown to enhance such correlations (5.1).

Test pits and shafts can be made quickly and cheaply, using a backhoe or a pier-drilling machine. The excavation depth of backhoe equipment is usually about 10–12 ft (3–3.7 m). Test shafts by a pier-drilling machine can go to any depth permitted by soil conditions, and are especially useful as drilled pier prototypes in defining caving or water inflow problems (see Section 5.9).

5.4.   The Standard Penetration Test (SPT)—"N-Value"

Soil samples from conventional test borings (in the United States) are usually taken with a 2-in. split-barrel sampler (ASTM Designation D-1586), combining the sampling with a penetration test of the undisturbed soil. The sampler is driven 18 in. into the bottom of the hole by blows of a 140-lb weight free falling 30 in. The number of hammer blows for each successive 6 in. of penetration is recorded; but not more than 100 blows if the 18-in. penetration is not achieved. The sum of the blows from the second and third 6 in. is called the penetration resistance, or the "N-value." This is a very useful test if performed carefully and with adequate equipment in good condition. Many design-related parameters—for example, pile-bearing loads, footing settlements, and so on—may be estimated from the "N" values and the boring logs, although these figures are quite sensitive to variations in operators' techniques and equipment suitability.

In one case in the authors' experience, the estimated pile lengths for the north half of a large building averaged 15 ft longer than those for the south half. A recheck on a few boring locations showed conditions to be fairly uniform over the building area. The difference in "N" values was traced to the different techniques of the two drillers used on the job: one habitually skimped on the 30-in. drop; the other, to be sure he gave good measure, always gave the cathead rope a little extra pull. The use of an automatic hammer-trip device would of course have prevented this variation, and it is a very desirable feature to have on a boring rig.

The foundation inspector who logs the test borings as they are made must monitor the driller's sampling technique and induce him to adhere to the ASTM method—which should be specified in the drilling contract.

5.5.   The Cone Penetration Test (CPT)

The cone penetration test (ASTM D 3441) (5.2, 5.3, 5.4, 5.5, 5.6, 5.7) is an instrumented sounding method that measures the resistance of soil deposits upon advancing vertically a 60°, 10-sq-cm cone at a rate of 10 to 20 millimeters per second. The rigid-jointed push rods have the same or smaller diameter as the cone for a length of at least 0.4 m above the cone base. The frictional force on a sleeve located just above the cone, the pore pressures generated during penetration, and the inclination of the tip may also be sensed and recorded. To eliminate rod–soil friction, the older mechanical cone penetrometers use a set of inner rods to activate a telescoping cone tip. Thus the cone is advanced to each test level by pushing in increments of 20 to 30 centimeters.

The newer systems incorporate electronic transducers integral with the cone and the friction sleeve. The electronic penetrometer is continuously advanced to refusal, and provides a complete record of cone penetration resistance as well as of any of the other parameters which may be sensed. The results of CPT measurements provide an enhanced definition of subsurface conditions, and can also be used to interpret soil strength and compressibility parameters (5.3) as well as empirical drilled pier design rules.

The cone penetration test, widely used in Europe since the 1920s, is beginning to find an increased following among geotechnical engineers in the United States, particularly in the exploration of cohesionless fill deposits. It is the authors' opinion that the general adoption of the CPT as a supplement to the use of SPT would result in more complete, accurate, and reliable geotechnical information in applicable geologic settings, for both bidding and design purposes. A typical cone penetration test record is shown in Fig. 5.1, and a photograph of a truck-mounted machine in Fig. 5.2.

5.6.   Other in-Situ Tests

In addition to the SPT and CPT, there are a variety of in-situ testing methods which can be used to derive soil and rock parameters in lieu of laboratory

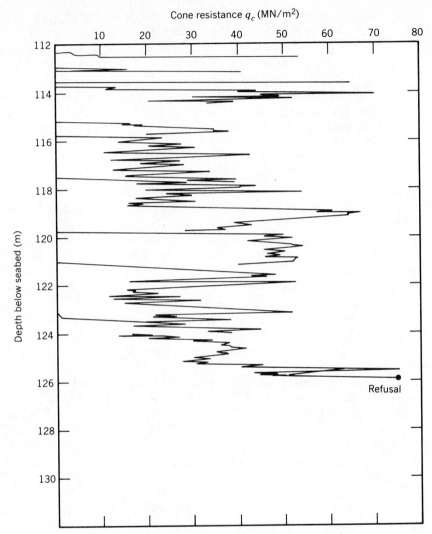

FIGURE 5.1.   Typical cone penetration record.

testing. Of these, the pressuremeter and dilatometer tests have the broadest
potential application in a variety of subsurface conditions. The pressuremeter
test (PMT) is made by inserting a cylindrical probe containing a water-filled
expandable membrane into a drill hole and applying pneumatic pressure
to a connecting reservoir. This causes a radial expansion of the membrane
against the side walls of the hole, and the volume change for each increment
of applied pressure is measured by the drop in water level displayed in a
manometer (5.8, 5.9, 5.10).

The dilatometer consists of a spade-shaped tip connected to a sounding
rod. The tip contains a small hydraulically activated flat cell which, at the

test depth, is expanded 1 mm against the soil in contact with the cell. An electronic load cell may be installed just above the spade tip to record the penetration resistance in a manner identical to that of the cone penetration test. The pressure recorded upon activation of the hydraulic pressure cell is interpreted to provide an estimate of the in-situ horizontal earth pressure, whereas the penetration resistance data can be interpreted in the same fashion as the CPT, allowing for the difference in the cone- and spade-shaped sounding tips (5.11).

5.7.   *Geophysical Methods for Locating Top of Bedrock*

Seismic traverses using refraction methods can be used to supplement test hole data in determining the depth to and the configuration of a buried rock surface, provided there is a marked difference in the seismic velocity of the rock and the overlying materials. Under favorable conditions, conventional multichannel seismographs will enable rock surface depth to be interpreted at 8 to 10 locations per day. The recorded rock velocities will also provide some insight into the quality (and drillability) of the rock. The seismic method, however, is often precluded in urban locations because of existing underground facilities and background vibrations.

FIGURE 5.2.   Cone penetration device, truck-mounted. (Photo courtesy of Triggs and Associates, Inc., Willoughby Hills, Ohio.)

Seismic reflection techniques using more sophisticated multichannel seismographs can provide an interpretation of detailed stratigraphic features including anomalies such as faults and fracture zones. However, this technique is typically not applicable above about 50 to 60 feet below the ground surface and is significantly more expensive than the refraction technique.

The resistivity method of exploration, most often using a variation of the four-pole Wenner configuration, has proven to have utility in definition of discontinuities within soils and rocks, for example, stratum changes, solution voids, major fractures, and so on. Interpretation of the apparent resistivity measured has been simplified by low-cost microcomputer processing and significantly increases the resolution and utility of this method. As with the seismic method, borings are required to provide a "calibration" of resistivity profiles and constant depth traverses.

A method for determining definitely when test borings have reached bedrock (as distinguished from a boulder or an isolated fragment of hard rock in a weathered zone) has been reported by Lundstrom and Sternberg (5.12). The method merits serious consideration.

A boring is drilled, definitely penetrating 10 ft or more into rock as confirmed by careful logging or rate of drilling, drill behavior, and so on (or by rock core). Normally, a percussion-type, air-operated rock drill is used; penetration is rapid, and this type of test boring is inexpensive. A hydrophone (a nondirectional underwater type of microphone) is set in the hole below rock level. Other test borings (or probes) are made at selected locations, and the sound transmitted during drilling is reproduced at the receiver–recorder connected to the hydrophone. It is reported that it is easy to distinguish, either from the sound or from the shape and amplitude of the recorded sound trace, between the drilling of continuous bedrock (either igneous or sedimentary) and of an isolated rock element, even when the latter is a very large boulder in contact with the surface of bedrock. An outstanding advantage of the method is that the test probes need only reach the top of sound rock, with no necessity for confirming by further penetration.

The radius of exploration for one hydrophone setting is said to be about 100 m (330 ft).

It appears that this method offers a relatively inexpensive, rapid, and accurate method of determining depth to sound bedrock for projects where foundations must rest on or in sound rock, and where overlying "hardpan," boulder till, or weathered rock make determination from penetration resistance alone difficult or expensive.

5.8.   *Large-Diameter Auger Borings for Exploration*

Because of the importance of being able to predict soil and rock behavior during pier drilling from the exploratory test results, it is often advantageous, and sometimes imperative, to drill prototype pier holes as a supplement to the conventional subsurface exploration. These test holes do not necessarily need to be the full diameter of the proposed pier holes; but they should be large enough that water entry, caving, sloughing, or squeezing in can be observed and logged accurately as to depth and stratum material and thickness. In some cases this can be done with a hole of only 12 or 16 in. in diameter (30–40 cm), logged from the surface; in others, particularly in holes of great depth or at sites where piers are to carry heavy loads, a complete and reliable log can be produced only by foot-by-foot inspection of the completed hole. This requires a boring with a diameter of at least 24 in. (60 cm), preferably larger, protective casing (in some cases), and all the safety precautions needed when anyone enters a pier hole.

It should be remembered that a poorly backfilled large-diameter test boring, if it is too close to a final pier location, can cause construction difficulties, or even unsatisfactory pier performance. For this reason it is prudent to limit test borings within the boundaries of the proposed structure to the conventional small-diameter variety, and to locate the large-diameter test borings outside the structure limits—which should be known and accurately located at that time. In any case, large-diameter test borings should be backfilled with care to forestall trouble of one kind or another at some future time. The use of a soil-cement slurry, a very lean concrete, or a mixture of dry sand with a little cement should be considered for backfill in any location where there is any possibility of later interference with pier-drilling operations or with distribution of loads from completed piers.

5.9. Exploration in Stony or Bouldery Soil

The use of large-diameter auger holes for exploration is of especial value and importance in bouldery deposits, such as are found throughout the world. Consider, for example, the following analysis. Suppose the site is one with scattered boulders, cobbles, and gravel—a glacial drift or till, for example. A large-diameter auger boring may be expected to bring up (though with some difficulty) a boulder about one-third the diameter of the auger; a 30-in. boring would bring up any stone up to 10 in. in maximum dimension, whose center lay within the 30-in. diameter, and be stopped by any larger stone encountered and by any 10-in. stone lying with more than half outside the auger's circle. This means that, so far as identification of 10-in. or larger stones is concerned, the 30-in. auger's circle of recognition is, not 30, but 40 in. in diameter, an area of 1250 sq in. Within this area, the auger will identify as being 10 in. or larger any stone that brings it to a stop.

Compare this with a conventional 4-in. test boring. It will bring up any small stone less than 1.3 in. in size (maybe), and be stopped by any larger than 1.5 to 2 inches. Any stone that stops it might be a boulder, a cobble, a large piece of gravel—or ledge rock. Its circle of recognition is about 4.5 in. in diameter and its area about 16 square inches—about 1.3%

of the area explored by the 30-in. auger hole; and the boring log cannot distinguish between boulders, cobbles, and large gravel.

In these circumstances, the small-diameter exploratory holes may mislead both designer and bidder; and there may be very little similarity between soil conditions depicted on the boring logs and those encountered when the pier shafts are drilled.

On the other hand, if the exploration is in a formation with only a substantial scattering of boulders, and few or no cobbles and gravel, a boring with a circle of influence of 1250 sq in. is much more likely to encounter at least some of the boulders than is a test boring that covers an area only 1.3% as large.

The authors have seen several "changed conditions" claims that would probably have been avoided if the exploration program had included a substantial number of large-diameter auger borings.

5.10. Scope of Exploration

The scope and type of exploration should reflect the type of support to be considered for the proposed structure. If there is a possibility of depending on sidewall shear in rock, for example, then obtaining rock cores for compressive testing could be important. If the site is in an area of known or suspected active clays, then frequent (or continuous) undisturbed samples of soils down to the depth of seasonal moisture content changes will be needed. The exploration program should be planned and carried out with such considerations in mind; and the boring contract should provide for its performance under competent inspection. It is not enough—it is not economical, and it may not be adequate—for the owner or his representative to specify something like "test borings on 50-ft centers, to a depth of 40 ft," and let a boring contract on that basis. Unless ground conditions are already very well known, the planning of a foundation investigation should be conducted by the geotechnical engineer in consultation with the architect and the structural engineer.

As a supplement to test borings, rock soundings using air-drilling methods have been found particularly advantageous at sites where depth to rock is extremely irregular; where there are potential solution cavities, or where the quality of the rock is variable, and where piers are planned as end-bearing on rock or as "rock sockets." A detailed conventional exploration of irregular or problem rock areas need not involve too great an extra expense, because usually most of the test borings can be sampled intermittently and can be supplemented by relatively inexpensive unsampled air-drilled holes. For some jobs, uncertainties can be reduced, and occasionally even substantial economies can be made by extending the exploration so that a conventional test boring or a sounding is made exactly at each column location.

Test borings should be continuously cored in rock (or drive-sampled in soft rock) for a sufficient depth to investigate all rock within the depth of significant stress increase. A record of the rate of advance (feet per minute, etc.) is also helpful in assessing the drillability of the rock, and microprocessors are available to provide automated recordings of the penetration rate. The detailed information on rock depth and hardness thus made available to prospective bidders for pier contracts will help avoid the need for substantial "contingencies" in the bid prices. The potential saving may be many times the cost of exploration.

A point requiring special attention in test borings where drilled pier foundations are a possibility is the possible need of sealing test holes after they are completed. Occasionally, where a pier location has coincided with the location of an earlier test boring, there will be water entry from a lower aquifer through the abandoned test hole, and the contractor will be unable to dewater the pier hole. This is a serious matter; at the very least, it will produce a delay in construction, and on several occasions it has led to the abandonment of a very large drilled pier and the substitution of another type of foundation (5.13, 5.14). When it is known that drilled piers may be considered, exploratory borings should be plugged from the bottom up to a level somewhat above possible pier bottom. Satisfactory plugs can be formed by dropping in, and tamping, bentonite pellets (sold commercially as Pi-pellets—see Appendix C), as is done for some types of piezometer; or the hole can be sealed by grouting from the bottom up. Grout for this purpose can be neat cement, cement–bentonite mix, or one of the chemical grouts. A grout that seals but does not offer much resistance to the auger is best.

In many geographical areas, particularly where piers are to be founded on rock, the general approach is to provide an estimate of base elevations solely to establish quantity estimates, whereas actual base elevations for each pier are determined by field conditions as interpreted by the foundation inspector. However, the subsurface exploration should be sufficient to enable interpretation of bearing contours which have a reasonable degree of reliability so that quantity estimates do not differ greatly from the as-built quantities.

5.11. Groundwater Observations

Not only should the level of water entry be noted on the exploration logs, but the circumstances and rate of water entry are of major importance in planning a drilled pier job. Each exploration hole should be used to obtain as complete coverage of this aspect as possible. The boring log should show the thickness and kind of formation yielding the water (e.g., sand and gravel overlying bedrock; weathered, broken, or solutioned rock; thin layers of clean sand within a clay formation; or fissured clay). Rate of water entry should be estimated or measured. On large projects where groundwater problems are anticipated, meaningful measurements can be made conveniently by using cased borings or piezometers with the tip seated at the level of a potential aquifer, by making pump-in (falling head) or pump-out (rising head) tests, or by measuring stable drawdown or pump-in level at a constant rate of pumping. Techniques and data treatment for field permeability testing of both rock and soil formations are summarized by the U.S. Bureau of Reclamation (5.15) and by Hvorslev (5.16).

A more recently applied variation of the rising-falling-head test involves dropping a plugged steel pipe (slug) in a test well, recording the maximum surge induced by the slug displacement, and recording (in real time) the subsequent fall from the maximum surge height to the equilibrium state. The change in water head during the test is measured by a submersible electronic transducer, is displayed on a strip chart, and may also be recorded on digital tape after signal processing. Note that by recording the maximum drop and the subsequent water level recovery induced by withdrawing the slug, both rising- and falling-head solutions for either confined or unconfined aquifers are obtained.

Seepage quantity estimates for full-scale pier holes made from pumping tests in small-diameter holes are of much more limited reliability than those made from large-diameter holes, but they can be very useful when larger-scale tests are not feasible, and may provide more specific data concerning individual aquifers.

5.12. Rock Coring and Quality

Whenever a rock formation is considered for pier support—either in end-bearing or in sidewall shear—the rock that is proposed for support should be investigated by core boring. Depth of coring will depend on the size of the pier or underream, the magnitude of the proposed load, and the nature of the rock formation. For end-bearing on rock, coring should extend below proposed bearing level to a depth of at least twice the bearing surface diameter. If poor recovery, or examination of recovered core, indicates the presence of voids, clay, or compressible zones within the interval cored, then coring should be extended further until a suitable continuity of sound rock is encountered. If the pier is to be supported by sidewall shear (a design that may be dictated by the condition of the rock penetrated), then depth of coring should be determined by the geotechnical engineer on the basis of conservative assumptions as to permissible shearing loads.

Good core borings in rock are expensive and require technical skill, experience, and good equipment. The authors believe that a negotiated contract with the best available drilling contractor is a better investment than letting a test-boring contract to the lowest bidder. By careful planning of the use of large-diameter auger holes (to allow preestimation of the penetration possible during actual drilling), correlated with carefully logged probes, core boring footage can be minimized.

Unless the rock to be cored is relatively sound and free from discontinuities that could influence core recovery, cores for drilled pier exploration should be of *NX size* (2⅜ in., or 60 mm) or larger, particularly where core compression testing is contemplated. As a general rule, the smaller the core diameter, the poorer the recovery. Often an *AX* or *BX* core will suggest poorer rock quality than actually exists, and may result in an initial estimate of deeper pier penetration into rock than is really necessary or economical.

To provide an assessment of the rock mass quality, the core recovered should be carefully logged by a qualified engineer or geologist noting: (1) rock type and hardness; (2) coloration and degree of weathering; and (3) attitude and condition of joints and fractures. Point-load tests (5.17) conducted on cores at the project site can provide uniaxial compressive strength correlations at significantly less cost than laboratory tests and should be considered for large projects.

Systems to rate the quality of the rock mass weight heavily the frequency of rock discontinuities and can be of assistance in evaluating pier drillability. The systems most frequently used include the Rock Quality Designation (RQD) (5.18) and the Rock Mass Rating (RMR) (5.19).

5.13. Rock Drillability

The correlation of conventional rock property data and small-scale test drilling (microbit tests) with the large-scale drillability of rocks in situ has long been the goal of researchers concerned with drilling for mineral and groundwater exploration and development, and is most recently being studied for application to tunneling by mechanical moles. It is unfortunate that almost no information on correlation of the penetration of large-diameter drills with material properties or with drill test parameters is available in the current literature. Thus, the application of such parameters to assess rock drillability for large-diameter drills is virtually unexplored, although some qualitative trends can certainly be established from the available small-bore research.

Perhaps the most definitive work to date involving small-bore drillability has been in the correlation of microbit drill tests with full-scale drilling, particularly with rolling-cutter bits. An example of the use of microbit testing to estimate the drillability of various rock types with a Tricone rock bit is shown by Fig. 5.3. For a comprehensive view of research applicable to small-bore drilling, the interested reader is referred to a report published by the Rock Mechanics and Explosives Research Center of the University of Missouri (5.20).

Most researchers have found that of all the rock properties considered, compressive strength of representative core samples is of the most value in predicting drillability. However, other factors important to rock drillability include tensile strength, abrasivity, rock fabric (grain size and orientation), residual stress, moisture content, rock structure, degree of weathering, and

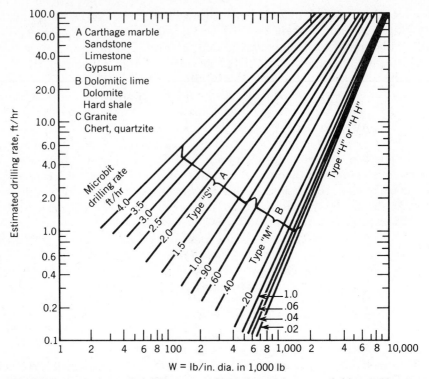

FIGURE 5.3. Estimated drilling rate of Hughes Tricone rock bits at 60 rpm as determined from microbit drilling test.

soft- and hard-rock interbedding. Current large-diameter drilling experience suggests that no one factor alone can be used to determine rock drillability. For example, a fine-grained crystalline rock with a heterogeneous granular orientation will drill much slower than a similar type of rock with coarse crystalline grains; and a rock with a high silica content (high abrasivity) but with a relatively low compressive strength would be very tough and wearing to drill. In contrast, a dense limestone or dolomite with a high compressive strength but a low hardness can be readily drilled.

Similarly, rock structure (fracture frequency, spacing, and the attitude and width of discontinuities such as joints, fractures, bedding planes, foliation, and schistosity) can control drillability. For example, joint spacings less than about 6 in. and greater than about 4 ft are usually favorable for drilling, whereas spacings in the range of 8 to 18 inches produce a type of blockiness that is difficult to drill. Bedding and layer thickness and depth may also significantly influence the drillability of the rock mass.

The influence of compressive strength on the drilling rate of various bit types is shown by Fig. 5.4. It is noted that for a given drilling thrust, the drilling rate generally increases with a decrease in compressive strength.

For the various types of drills and drill bits, drillability of rock formations is dependent on many factors which are related to such tool variables as speed of rotation, thrust, and torque, which in turn are related only indirectly to rock properties. It may therefore not be possible to derive simple, representative drillability relationships between drill characteristics and rock properties, particularly for large-diameter borings. However, even rough qualitative correlations between drillability and simple, small-scale test parameters are urgently needed to permit a more rational assessment of drill selection and of the economic feasibility of large-diameter rock drilling. Such assessments would go far to improve the confidence of the engineer and contractor in the use of piers drilled into hard-rock formations to sustain the large loads that are now being imposed with increasing frequency in urban construction.

A recent innovation in rock drilling is the synthetic-diamond drill introduced in 1979. The polycrystalline diamond is bonded to tungsten carbide to form cutters that bore perpendicularly into the rock. These bits made in the United States by Christensen Diamond, N. L. Hycalog, American

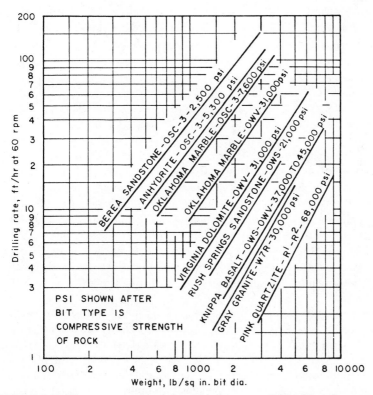

FIGURE 5.4.   Drilling rate versus weight on bit for new rock bits. Note that correlation with compressive strength of rock is irregular, but shows a general trend.

Coldset, Strat Bit, and Davis and Hicks are said to outproduce the conventional tricone rock bits except in the harder abrasive rock. It is likely that such bits will be adapted to pier drilling if such has not already occurred.

References

5.1   Pfister, Paul, Recording Drilling Parameters in Ground Engineering, *Ground Eng.*, **18**(3), (April 1985).

5.2   ASTM Designation D 3441-79, "Standard Method for Deep, Quasi-Static, Cone and Friction-Cone Penetration Tests of Soil."

5.3   Triggs, J. Fred, Jr., Dutch Cone Penetration, an Alternative to the Test Boring, *Foundation Drilling*, May 1984, p. 10.

5.4   Begemann, H. K. S. P., The Friction Jacket as an Aid in Determining Soil *Proc. 6th Intern. Conf. SM & FE, Montreal*, 1965.

5.5   Begemann, H. K. S. P., The Use of the Static Soil Penetrometer in Holland, *N. Z. Eng.*, February 15, 1963.

5.6   Sanglerat, G., *The Penetrometer and Soil Testing*, Elsevier Publishing Company, New York, 1972.

5.7   Robertson, P. K., and Campenella, R. G. Interpretation of Cone Penetration Tests—Part I: Sand; Part II: Clay, *Can. Geotech. J.*, **20**(4), (November 1983).

5.8   The Menard Pressuremeter, *Sols Soils*, No. 26 (1975) (Paris, France).

5.9   Lucas, R. G., and de Bussy, B. LaC. Pressuremeter and Laboratory Test Correlations for Clay, *ASCE J. Geotech. Eng. Div.*, Vol. 102, No. GT9; Proc. Paper 12410, September 1976, pp. 945–962.

5.10 Baguelin, F., Jezequel, J. F. and Shields, D. H. *The Pressuremeter and Foundation Engineering*, Transtech Publications, Clausthal, Germany.

5.11 Marchetti, S. A New In-situ Test for the Measurement of Horizontal Soil Deformability, *Proc. ASCE Specialty Conf. on In-situ Measurement of Soil Properties*, Vol. I, 1975, Raleigh, North Carolina.

5.12 Lundstrom, R., and Sternberg, R. Soil-rock Drilling and Rock Locating by Rock Indicator, *Proc. 6th Intern. Conf. SM & FE, Montreal*, 1965.

5.13 Tomlinson, J. M. Discussion of Paper by Palmer and Holland, Session B, Large Bored Piles, *Proc. Symp. Inst. Civil Eng., London*, 1966.

5.14 Toms, A. H. Discussion of Session B, Large Bored Piles, *Proc. Symp. Inst. Civil Eng., London*, 1966. (See also closure of same discussion by D. J. Palmer.)

5.15 U.S. Bureau of Reclamation, "Earth Manual," Designations E-18, E-19, E-36.

5.16 Hvorslev, M. J. "Time Lag and Soil Permeability in Groundwater Measurements," Corps Eng., *Waterways Expt. Sta., Vicksburg, Miss., Bull.* 36, 1951.

5.17 Bieniawski, Z. T., and Franklin, J. E. Suggested Methods for Determining the Uniaxial Compressive Strength of Rock Materials and the Point-load Strength Index, *Commission on Standardization of Laboratory and Field Tests, International Society for Rock Mechanics*, Committee on Laboratory Tests, Document No. 1, 1972.

5.18   Deere, D. U. et al. Design of Subsurface and Near-surface Construction in Rock, *8th Symp. on Rock Mechanics, Proc. AIME*, 1967.

5.19   Bieniawski, Z. T. Determining Rock Mass Deformability—Experience from Case Histories, *Int. J. Rock Mech. Min. Sci.*, **15**(51), (1978).

5.20   "Rock Properties to Rapid Excavation," report by Rock Mech. Expl. Res. Center, University of Missouri, Rolla, MO (U.S. Dept. of Transportation Contract 3-0143).

6

Drilled Pier Specifications and Contract Documents

6.1. Introduction—General Considerations

The specifications and contract for a drilled pier project should give the contractor as much leeway as practicable in the choice of equipment and methods. A performance type of specification, imposing no unnecessary restrictions but requiring an acceptable end result, will in the long run result in more competitive bids, lower bid prices, fewer disputes and claims for extras, and usually better foundations than would a tight specification that unnecessarily restricts the choice of contractors or impedes the work of the successful bidder. For example, if not required by a local building code, specification of a 60° bell (see Fig. 1.1) might eliminate bidders having only equipment for 45° bells or dome-shaped bells, which are usually just as good; or unnecessary specification of a minimum torque for the drilling machine could discourage bidders with slightly smaller machines who had learned to drill the specified sizes of holes or bells in the site formations by the use of special techniques or special drilling tools. There will be occasions when it is necessary to limit the contractor in such matters, but the limitations should not be imposed unless the need is real and important.

A specification that is "too tight"—that is, which imposes unnecessarily difficult or restrictive conditions on the contractor—may lead to bids which are unnecessarily high, but more probably will result in some of the bidders ignoring the restrictive provisions entirely on the theory that "they can't possibly be enforced; so why include them in cost considerations?" And when the successful bidder begins to ignore requirements that appear unreasonable, the owner's representatives must either enforce the specifications, with wrangling and risk of lawsuit, or else waive the objectionable requirement—a dangerous precedent for the job, and certainly a concession that is unfair to the other bidders.

One difference between well-drawn plans and specifications for a drilled pier job and those for a more conventional construction project is that the drilled pier plans must be more flexible, for underground conditions must always be inferred from limited data, and the exact conditions that will be encountered at a site will usually remain unknown until the last pier hole is drilled. This means that there must be provisions for field decisions to found piers deeper (or shallower) than planned, or to change shaft or bell sizes, or to place additional piers when required by deviations in pier location or plumbness. And in some geologic situations economy may dictate a change in pier type as the work proceeds and the underground picture becomes clearer.

The plans, the specifications, and the contract documents should reflect the need for this field engineering. The needed flexibility will vary from a minimum for sites where the geologic formations are regular, uniform, and well known, to a maximum for sites where heavy loads or important structures are to be carried by irregular deposits or by decayed, fissured, cavernous, or pinnacled rock. The project specifications should be consistent with what is known—and what is unknown—about site subsurface conditions.

Drilled pier plans and specifications are sometimes prepared by engineers or architects who are not thoroughly familiar with the problems and techniques of drilled pier construction, with local geology, with what is known about site conditions, or with local pier-drilling practice and equipment. This can lead to misunderstandings and complete disagreement between owner, engineer, and contractor, with consequent job delays and cost increases. Attempts by an inspector or engineer to enforce compliance with vague or inappropriate requirements will usually result in weakening of the authority of the engineer, relaxation of the project specifications, and ultimate claims for "extras."

Specifications should not be prepared on a "rush" basis. When drafted, they should be reviewed by the project geotechnical engineer and by experienced legal counsel, with especial attention to clarity, completeness, consistency, and the presence of terms that could be misinterpreted.

It is neither prudent nor economical to use a standard set of specifications for all drilled pier jobs, changing only the pertinent dimensions and numbers. The specifications for each job should be written to provide for all contingencies that are predictable on the basis of available site information, and should allow as well for the possibility of unforeseen field changes. This means that each set of specifications should be carefully tailored to the job at hand. Pay items designated in the bid documents and contract should cover all probable work items, so as to allow the contractor to get a fair price for all his work without having to ask for "extras" or claim "changed conditions." Care should be taken to avoid unrealistic quantity estimates for any pay item that affects the bid total, for such errors are an invitation to unbalanced bids from contractors who recognize the error because of their local experience. Typical of this sort of error is an underestimate of rock quantity. In some geological circumstances, the term "rock" (a pay item) may be questionable—or arguable. There may be hard and soft layers; strata that cannot be drilled without using a core barrel or other rock bit, alternating with strata that can be drilled with an ordinary earth auger. Because the latter cannot be cored readily, such a sequence will require switching back and forth from one drilling tool to another, a time-consuming and labor-intensive activity as costly to the contractor as continuous rock drilling. This has led to the inclusion in the definition of rock in the current ADSC specifications (see Appendix C) of the following sentence: "All earth seams, rock fragments, and voids included in the rock excavation area will be considered rock for the full volume of the shaft from the initial contact with rock for pay purposes."

The authors know of one court case in which the ADSC rock definition was cited by the Court in its decision (6.1). Wherever the foundation investigation (or general knowledge of local geologic conditions) indicates the possibility of anomalous soil/rock classifications, a suitable definition of rock for pay purposes should be included in the specifications.

For sites located in areas where ground conditions cannot be predicted with certainty, the bid documents should state that quantity estimates are for bid purposes only, and should call for over and under unit prices, with a statement that bids will be rejected if they contain unit prices that are, in the judgment of the owner's representative, unbalanced.

The authors have seen disastrous cost increases where these precautions in the bidding documents had been neglected. Because of these possibilities, the bid and contract documents, as well as the plans and specifications, should be prepared or reviewed by a geotechnical engineer experienced in all phases of design and construction of drilled piers, and acquainted with site geologic conditions, before bids are advertised.

Bid documents should also require bidders to submit a list of projects in similar subsoil and groundwater conditions that they have completed successfully. For projects of special difficulty, or where proven capabilities and experience are essential, prequalifying prospective contractors and negotiation of contracts in lieu of competitive bidding is recommended to provide the owner with the best and least expensive construction.

As has been stated elsewhere, the designer of the structure is normally responsible for redesigns made necessary by ground conditions which result in deviations of piers from position, alignment, or plumbness. The designer is usually designated in the contract documents as the structural engineer; or the terms "engineer of design," "engineer of record," or "architect/engineer" may be used. In the discussion that follows, and in the sample specification (Subsection 6.3), the term "structural engineer" will be used for this agency, and the terms "geotechnical engineer" and "foundation inspector" will be used as defined in Chapter 4.

The scope and the limitations of the duties that go with these positions should be set forth clearly in the contracts between the owner (or his agent) and these several agencies. They may be included also in the construction contract, or may be incorporated therein by reference to the special contracts concerned. In either case, the contractor should be enabled to acquaint himself with the duties and limitations of the engineering/inspection agencies that he will have to deal with.

Contracts often require the contractor to furnish inspection services as part of his contract. The object is, of course, to enable the owner to include the cost of inspection as part of the construction cost when applying for a construction loan or for a mortgage. This practice is wrong; it is against the interests of both owner and lender. Any inspection agency *must* represent the interests of the owner. An inspection agency which is paid by the contractor can be subjected to conflict of interest, no matter how good its intentions or how firm its resolve. An inspection decision that costs the contractor money may be protested; disputes may develop; and the contractor, resentful at being required to pay for what he considers "bad" decisions, is in a position to express his resentment by holding back, or even refusing, payments to the inspection agency. This is not merely a hypothetical possibility. It happens on many contracts, and it benefits nobody. An inspection agency should be engaged by, report to, and be paid by the owner or one of his agents. Lending agencies would find it to their interest to recognize this situation and allow inclusion of inspection costs as part of construction costs.

The presentation of soil or rock profiles as part of the plans is a particularly risky procedure. Any such presentation will normally be prepared by a draftsman, usually without training in geology or soil and foundation engineering. Errors in wording or symbols are common. It is imperative that all subsurface data presented in the plans be checked in every detail by the geotechnical engineer, both to be sure that all data have been correctly presented and to be sure that nothing in the presentation might be taken as fact when it is no more than inference.

All of the subsurface information available to owner, architect, structural engineer, or geotechnical engineer should be made available to prospective bidders; and this should include the soil and rock samples from the subsurface exploration, preserved at natural moisture content and arranged so that they can be readily examined. The samples, of course, will have to be kept at some central location, such as the geotechnical engineer's or the architect's office. The geotechnical report (or soil and foundation report), with complete boring logs, should be reproduced and made available to prospective bidders. Making these documents available for examination in the architect's office is not sufficient. A contractor needs to be able to study them freely and at length while preparing his bid.

To be complete, a specification for drilled pier construction should cover many items, both general and specific for the project covered. It is easy to overlook and fail to cover an item which can come up later as the basis for a question, or a dispute, or even a claim for an "extra." To help the reader check his specifications for completeness, the following guide is offered. It should not be expected to cover all possible contingencies. New structural requirements, new construction techniques, variations in ground conditions—all these things require variations in specifications. Perhaps the use of the guide will suggest some of these variant items.

6.2.   A Guide to Technical Specifications for Drilled Piers (Shafts) (Caissons)

6.2.1.   General Conditions

1. Definitions. The Contractor is defined as the organization responsible for drilled pier (shaft) installation work. The Engineer is defined as the individual or organization who represents the Owner during the drilled pier (shaft) installation work. The Engineer will be identified by the Owner prior to the commencement of the work.

2. Subsurface conditions. The Contractor shall familiarize himself with the site and subsurface conditions as represented by Contract Drawing Nos. _____ and Contract Document No. _____ . Detailed subsurface information records are on file in the office of _____ .

3. Qualifications and requirements. Drilled piers (shafts) shall be installed by a Contractor experienced in drilled shaft work. The shaft excavations shall be performed with an approved drilling rig capable of expeditiously performing the work specified herein. Casings shall be maintained in good shape and free from old concrete on the inside surfaces that would tend to make pulling difficult. All equipment shall be maintained in satisfactory working condition and shall be operated by competent and experienced personnel.

4. Site cleanup. Upon completion and acceptance of all specified work, the Contractor shall promptly remove from the site all unused construction materials and equipment. The work area shall be subsequently restored to the satisfaction of the Engineer.

6.2.2.   Scope of Work

1. The work specified herein shall consist of furnishing all labor, equipment, and services necessary for, and reasonably incidental to, the construction of drilled piers (shafts) shown on Contract Drawing No. _____ and in strict accordance with the terms and conditions of the Contract Documents.

2. Exclusions. Excluded from this section is the field engineering layout, including all lines and grades required for the drilled pier construction.

6.2.3.   Shaft Excavation

1. Excavation method. Piers (shafts) shall be drilled through overburden and _____ in accordance with the lines and grades shown on Drawings No. _____ . Excavation shall be advanced in a manner which will not adversely affect the integrity and performance of the completed piers or damage adjacent facilities and property. Shafts shall not be excavated within 10 ft of previously completed piers until the concrete has been allowed to cure for at least seven days.

2. Liner casing. Unless a written exception is granted by the Engineer, temporary steel casing or an acceptable equivalent shall be provided for the full depth of the pier (shaft) excavation. The casing shall be of sufficient strength to sustain handling and the pressures imposed by earth, fluids, and concrete during construction. The casing shall not be out-of-round by more than two inches. Casing sections with cutting teeth or equivalent shall be provided as required to drill casing a limited distance into hard soil or soft rock.

3. Grooving. If required by the design provisions, a series of grooves shall be cut into the side walls of the rock socket at vertical intervals of not more than 8 in. (20 cm). The width and depth of the grooves shall be not less than ¼ in. (64 mm). The grooves may be continuous (spiral), or unconnected (rings).

4. Water control. Shaft excavations shall be maintained in an essentially dry condition, by pumping if necessary, until just prior to concreting. Alternatively, slurry drilling procedures which are acceptable to the Engineer may be used if water infiltration is excessive as defined by Subsection 6.2.6-2.

Dewatering operations shall be conducted in such a manner so as to prevent ground loss in shaft excavations and surface subsidence due to ground loss or subsoil consolidation.

5. Control of bentonite slurry. Bentonite shall be mixed with clean, fresh water to produce a slurry sufficient to maintain the stability of a shaft drilled in potentially caving and/or water-bearing soils. The bentonite concentration shall normally be between 4 and 8% by volume and the slurry shall conform to the following provisions.

Properties of bentonite. Bentonite shall be generally in accordance with API Specification RP13B. The manufacturer's certification of the bentonite specifications shall be provided to the Engineer, upon request.

Properties of slurry. The following tests shall be applied to the field-mixed bentonite slurry. The results of the tests shall be within the ranges tabulated below unless otherwise approved by the Engineer.

Slurry Properties	Units	Acceptable Range (20°C)	Test Method[a]
Density	gm/cc	1.03–1.08	Mud density balance
Viscosity	sec	30–50	Marsh cone
	or cP	3–10	Fann viscometer[c]
Shear strength[b]	N/m	2–14	Shearometer
pH		9–12	pH Indicator

[a] API Specification RPI3B.
[b] Ten-minute gel strength.
[c] Fluid sample screened by #52 sieve.

Test frequency. The density of the bentonite slurry shall be measured daily and shall include, prior to concrete placement, tests at a depth of about six (6) feet above the base of the shaft to check for slurry contamination. Densities so determined shall not exceed 1.20 g per cc. Other tests shall be carried out until a consistent working pattern has been established taking into account any changes in the slurry due to blending of freshly mixed suspension, re-use of slurry, or processing to remove impurities. Additional tests for shear strength and pH shall be made from time to time if required by the Engineer.

Mixing water. The temperature of the water used in mixing the bentonite shall not be less than 5°C. If saline or chemically contaminated groundwater is encountered, special measures shall be taken to modify the slurry subject to approval of the Engineer.

6. Bearing surface preparation. Unless otherwise specified, the end-bearing surface shall be cleaned of all loose or softened materials, debris or other substances sufficient, in the Engineer's opinion, to cause settlement or to affect concrete strength.

For rock surfaces with an average dip greater than one (1) vertical to one and one-half (1.5) horizontal, the rock surface shall be stepped, with the maximum step height less than one-quarter of the shaft-base diameter. Alternatively, steel dowels grouted in the rock shall be provided to the satisfaction of the Engineer.

7. Rock excavation classification. Unless otherwise specified, rock is defined as material which cannot be drilled with a heavy-duty auger drill employing a toothed auger designed to excavate hard soil and soft or weathered rock.

6.2.4.   *Shaft Dimensions and Alignment*

1. Shaft dimensions. The shaft shall be installed to the dimensions specified on the Contract Drawings and shall be carried to the bearing levels determined in the field for each pier by the Engineer.

2. Shaft location and grade. Completed shafts shall be centered within 1/24th of the diameter but not more than 3 in. from plan location. The cutoff elevation of such shafts shall be within ± 3 in. of the established cutoff elevation.

3. Shaft alignment. A shaft shall neither deviate from the vertical by more than one and one-half (1½) percent of its length, nor exceed an offset more than twelve and one-half (12½) percent of the shaft diameter or fifteen (15) inches (38 cm), whichever is less. Battered shafts shall not deviate from the prescribed alignment by more than five (5) percent of the shaft length. Shaft alignment deviation in percent is defined as the horizontal projection of the distance between the centers of the shaft at the base and at the cutoff elevation divided by the vertical length of the shaft times 100.

4. Out-of-tolerance-shafts. If the tolerances specified under Subsections 6.2.4-2 and 6.2.4-3 above are exceeded, the extent of overloading produced shall be investigated. If in the judgment of the Structural Engineer corrective measures are required, the Contractor shall implement such measures at his own expense.

6.2.5.   Reinforcing Steel

1. Placement. After completion of the excavation, steel reinforcement, if specified, shall be placed in each shaft excavation with provisions to ensure that the reinforcement is centered in the shaft and is so maintained during concrete placement. Unless otherwise directed by the Engineer, reinforcing steel supports shall extend to the base of the shaft. During extraction of casing, the upward and downward movement of the top of the reinforcement shall not exceed 5 in.

2. Materials. Reinforcing steel shall meet the requirements of ASTM A 615 or A 706 and shall be free of mud, grease, oil, or other surface coatings which may impair the steel–concrete bond. Bars containing rust and scale may be used provided wire-brushed specimens meet the specified dimensions and weight of the bars.

3. Design. Reinforcing cages shall be designed to be stable and to retain the configuration of the bars during concrete placement. The clear distance between bars shall be at least one and one-half (1½) times the maximum aggregate size. Horizontal reinforcement of cages shall be either spiral caging or horizontal loops.

6.2.6.   Concrete Placement

1. Free-fall. After the bearing surface is prepared and reinforcement placed, concrete shall be placed in a manner that will preclude segregation of particles, excessive infiltration of water, or any other occurrences which would tend to decrease the strength of the concrete or the supporting capacity of the finished caisson. Concrete shall be placed through a center hopper, or equivalent, in such a manner as to free-fall vertically without obstruction.

2. Tremie. If free-fall cannot be assured without obstruction or if shaft excavations up to six (6) feet (1.8 m) in diameter contain more than three

(3) inches (7.5 cm) of water, concrete shall be placed by the tremie method or by pumping in place. Limiting water depths for larger-diameter shafts or bells shall be determined by the Engineer. Tremie pipes shall be preferably twelve (12) inches (30 cm) and not less than eight (8) inches (20 cm) in diameter. Positive control shall be provided to ensure that the bottom of the tremie pipe is at all times below the concrete surface. Before placement of concrete by the tremie method, the shaft shall be filled with water to the natural groundwater level so that the water heads inside and outside the shaft are balanced.

2. *Casing withdrawal.* During withdrawal of the casing used to retain the excavation during concrete placement, the column of concrete in the casing shall at all times be higher than the head of groundwater or trapped slurry (whichever is highest) acting at the bottom of the casing, but never less than five (5) feet (1.5 m). The method used to pull the casing and to document the relative head levels will be subject to the approval of the engineer. Concrete levels shall be checked before, during, and afer the casing is pulled.

If casing set in predrilled holes is not removed, the engineer may direct that the annulus between the casing and the soil/rock be pressure grouted with an approved cement grout, or otherwise filled so as to restore a tight contact between the casing and the soil/rock.

6.2.7.   *Concrete Properties*

1. *General requirements.* Concrete shall have a 28-day strength of at least _____* psi and shall meet the relevant requirements of "Specifications for Structural Concrete for Buildings," ACI 301, unless otherwise noted herein. Concrete shall not be used which has had water added to the mix more than 1 hr before placement unless otherwise approved by the Engineer.

2. Concrete mix. The mix design and documentation of the concrete strength shall be submitted to the Engineer for approval prior to initiation of the work. For concrete placed by conventional means, the coarse aggregate size should not exceed one (1) inch (2.5 cm) for plain concrete or lightly reinforced shafts nor ¾ in. (2 cm) for heavily reinforced piers. For these cases, the slump of the concrete shall be five (5) inches ± one (1) inch (125 mm ± 25 mm) and six (6) inches ± one (1) inch (150 mm ± 25 mm), respectively.

Specially designed mixes shall be used for tremied and pumped concrete and shall be submitted to the Engineer for approval before use. The slump of tremied and pumped concrete shall be at least six (6) and seven (7) inches (15 and 18 cm) respectively, with no increase in water–cement ratio.

---

* To be determined by the Structural Engineer, but not less than 4000 psi (280 kgf/cm^2).

6.2.8. Inspection and Testing

1. Inspection. The Engineer will provide continuous on-site inspection during the duration of the pier installation. The Contractor shall cooperate with the Engineer during the inspection process and assist in securing the construction documentation specified herein.

2. Testing of bearing materials

Rock testing. Testing of rock shall consist of recording the rate of penetration of a pneumatically activated one-and-one-half (1½) to two-(2)-inch (3.8−4 cm) diameter steel star-bit. The depth of penetration shall be equal to the base diameter of the shaft or bell or to a depth of six (6) feet (1.8 m), whichever is greater. The drilling shall be accomplished under a uniform drill pressure in a manner to prevent binding of the drill bit. Permanent records of the time required for each six (6) inches (152 mm) of drill penetration shall be recorded at each test location. The depth and extent of any fractured zones, seams, or voids logged shall also be recorded for permanent record.

The Engineer may probe all test holes with a "feeler" rod to investigate the extent of any fractured zones, seams, or voids. Should the results of the penetration testing, in the Engineer's opinion, be inconclusive, the Engineer may direct that NX-size rock cores of the bearing materials be obtained. The Contractor shall provide whatever assistance is required to facilitate any core drilling operations.

Soil testing. Unless otherwise specified, testing of bearing soils will be conducted by the Engineer. Should the Engineer require test borings to provide samples of the bearing materials, the Contractor shall provide whatever assistance is required to facilitate any test drilling operations. If the pier has been designed to develop a portion of its capacity by skin friction, the Contractor shall provide whatever assistance is required, including partial casing withdrawal, to expose the side walls for inspection and testing.

3. Concrete testing. One set of four cylinders per each 100 cu yd (75 m³) of concrete placed but not less than one set per pier shall be taken for concrete testing. One cylinder shall be tested at seven days and two at 28 days. One sample shall be kept in reserve for testing in the event of a low break. The concrete sampling shall be done in accordance with ASTM C 172 and the cylinders shall be prepared and cured in accordance with ASTM C 31. The slump of concrete shall be checked in accordance with ASTM C 143 to satisfy the requirements of Section 7.02 above.

4. Safety. The Contractor shall be responsible for the safety of all personnel entering shaft excavations for construction and inspection purposes. Safety equipment shall be provided and shall include gas-sensing equipment, protective cage or equivalent provisions and, if required, facilities for forced air ventilation of shafts, as well as other equipment or facilities required by local, state, or federal regulations.

Before entry, shafts shall be checked for toxic and explosive gases. If such gases are encountered, forced air shall be circulated in the shaft until safe entry conditions are certified or adequate safety measures are taken to the satisfaction of a health and safety specialist.

6.2.9. Records

The Contractor shall maintain for each shaft complete records of drilling, reinforcement, and concrete placement, and other operations required under this contract. These records shall be submitted to the Engineer on a daily basis. For each shaft, the records shall include, but shall not be limited to:

1. Detailed description of drilling equipment and tools used.
2. Daily production log.
3. Shaft locations, alignment, and dimensions.
4. Reinforcing steel lengths and cutoff elevations.
5. Concrete volume placed.
6. Total rock drilling depth and drilling time.
7. Pay item quantities.

Within one week of the termination of the drilled shaft construction, a summary report shall be submitted by the Contractor containing all pertinent records and other data relevant to the work.

Supplementing the preceding guide, Appendix C in this book presents the current (1983) specification of the ADSC, which incorporates the provisions of the American Concrete Institute's Specification ACI 336.1 with additions by ADSC. At the time of this writing (1985), committees of the ADSC and the ACI are working on an updated specification, and it is anticipated that the new one will be available, perhaps by the time this book is published.

Also of interest is a British specification for cast-in-place piles formed under bentonite suspension (6.2).

References

6.1   Parks Drilling Company *vs.* City of Akron, No. 76-7-1476 (Summit Co., Ohio, Court of Common Pleas, undated).

6.2   Specification for Cast in Place Piles Formed Under Bentonite Suspension, *Ground Eng.*, **18**(4), 11–15 (May 1985). (This document can be obtained from the Federation of Piling Specialists Ltd. of Dickens House, 15 Tooks Court, London EC4A 1LA, England.)

7

Minimizing Construction Claims—Excessive Costs—Liabilities

Foundation construction is probably more subject to cost overruns, claims for "extras," and budgeting uncertainties than any other phase of building construction. The deeper the foundations, the more troublesome these cost uncertainties are likely to become; and drilled pier foundations are probably more subject to unanticipated costs than most other types of deep foundations. The reasons for this are inherent in subsurface variations: soil and rock conditions can be significantly variable within very short distances; the knowledge of subsurface conditions available to the contracting parties is always limited and imperfect; and the techniques of pier drilling and concreting are sometimes more sensitive to variations in soil, rock, and groundwater conditions—as well as to the effects of bad weather—than, for example, pile-driving operations. When unanticipated costs or claims develop, they may involve large sums of money, sometimes more than doubling the cost of the subsurface phase of a project. They may delay (or even prevent) the completion of construction; and they can be very damaging to the economics of the project, to the contractor's reputation, and to the engineer's or architect's relations with his client. The threat of major cost increases may require redesign, or reduction of the scope of a project, or may involve difficult revision of financing arrangements. If cost overruns occur frequently in the practice of an engineer or architect—or in the performance of a contractor, or in construction, generally, in a particular area or region—they can adversely affect the availability of insurance coverage.

Such cost increases and claims cannot be eliminated completely, but they can be minimized, if their causes are understood and their avoidance is included in planning and in contracting.

7.1. "Changed" Conditions

This term is used to designate conditions which affect the cost of construction and which are different from the conditions described in the plans and specifications, the contract, or any of the supporting data or information made available to the contractor. The term is not usually intended to imply that conditions have changed from what they once were but rather that they are different from those originally represented. "Changed conditions" clauses are put into specifications with the hope that bidders will feel that their risk of loss due to adverse subsurface conditions is covered and that they can reduce or eliminate the contingency allowance in their bids. Most commonly, "changed conditions" are described as in the following example taken from a set of specifications for a major project:

> *Changed conditions*—The Contractor shall notify the Engineer in writing of the following conditions, hereinafter called changed conditions, promptly upon their discovery and before they are disturbed:
>
> (1) Subsurface or latent physical conditions at the site of the work differing materially from those represented in this contract: and
> (2) Unknown physical conditions at the site of the work of an unusual nature differing materially from those ordinarily encountered and generally recognized as inherent in work of the character provided for in this contract.

"Changed conditions" in this sense, if proved, often constitute the basis for settlement or court awards to contractors. It is to the advantage of everyone concerned that construction conditions be known as completely as practicable before plans are prepared or bids are taken.

To this end, subsurface exploration programs should be planned with both the requirements of the structure and the probable subsurface conditions of the site in mind, and the man in charge of exploration should expect to modify his plans, if necessary, as exploration proceeds. It does not make sense to lay out in advance an array of borings to preassigned depths, without regard to what may be encountered below the surface when the borings are made. If a "changed conditions" claim is not to be encouraged, the exploration program must be adequate, both in horizontal coverage and in depth of borings, and must be reported in detail and in terms that contractors can understand. Moreover, samples of the formations penetrated must be preserved at natural moisture content for examination by prospective bidders. A discussion of exploration for drilled pier projects will be found in Chapter 5.

The best prevention for "changed conditions" claims is to make certain that you "tell it as it is" to the prospective bidder on your pier job. We can state from experience that some of the most common grounds for "changed conditions" claims on drilled pier construction lie in the areas listed below, when the contractor was uninformed or misinformed:

1. Type of rock or obstruction encountered (limestone, sandstone, shale, boulders, cobbles, "hardpan").
2. Accuracy of depth of rock layers.
3. Variation between hard and soft layers.
4. Stability of unconsolidated material to be drilled.
5. Presence of squeezing or caving layers.
6. Depth to water table and probable seasonal variations.
7. Permeability of water-bearing layers.
8. Presence of artesian pressures.
9. Structure and hardness of rock to be drilled (joints, fracturing, dip, foliation, thickness of weathered zone).
10. Frequency, size, and shape of cobbles or boulders encountered.
11. Unexpectedly difficult access.
12. Excessive overdrilling (for various reasons, but sometimes due to in-adequate—or inexpert—inspection).

The contractor must estimate his costs from what the plans, specifications, and other documents furnished tell him or do not tell him. These must be accurate and unambiguous. Most subsurface conditions tend to be hetero-geneous rather than uniform, and usually subsurface reports and specifications are based on relatively few small-diameter borings for jobs requiring many drilled piers. It is important that in addition to presentation of the logs of the borings, the soil and foundation report should include results of ob-servation of geologic conditions at and at some distance surrounding the site, in order to give the contractor as complete a picture as possible of what to expect. For instance, the fact that small borings do not encounter cobbles or boulders does not necessarily mean they are not there, especially in a type of sedimentary formation that commonly includes cobbles. Also it is not enough to say merely that cobbles and boulders are present. Their size and shape and frequency should be predicted if at all possible.

It would be prudent for the engineer to include some large-diameter borings in the geotechnical investigation of a site where there appears to be a possibility of extensive use of drilled piers. Although this may increase the cost of the investigation, it also may avoid delays and troublesome claims during or after construction.

If water is not present in a test boring, it is not sufficient to omit a water table symbol on the log. There should be a specific statement that it was absent and at what time of the year, and whether there might be water during another season.

These are only a few illustrations of the sort of "changed conditions" pitfalls that can be encountered. The engineer who is trying to avoid "changed conditions" claims must be willing to tell the bidder that certain conditions are unknown to him, if this is true, and must remember to put the responsibility upon the bidder to find the answer or make his own interpretation before he bids. And—very important—he must give the bidder ample time and opportunity to make his own study.

A last point of advice to both engineer and contractor on a large pier-drilling operation, especially a troublesome job, is that good documentation and recordkeeping are most important. When difficulties are encountered by a contractor, the contractor's superintendent should make a *complete* record in his diary; and the inspector or the geotechnical engineer should record the nature of the difficulty, the contractor's methods in handling the problem, the type and condition of the equipment being used, the estimated time lost, and his opinion as to the cause of the difficulty. Photographic records of the situation can be of real value in settling claims before they become formal, or in arriving at an equitable settlement in case claims subsequently arise.

Although design changes during construction are sometimes considered to *constitute* "changed conditions," they are more commonly used as *evidence* of "changed conditions." Most design changes are covered under a special paragraph in specifications and generally make negotiation of revised payment necessary if the change is significant.

7.2. Extras

Cost increases due to "extras"—items not anticipated or priced in the bid documents or contract—are almost inevitable on a large project, but they can be minimized by planning and by alertness on the part of all concerned as the construction proceeds. If there is a possibility of a new cost item being incurred, although it is not expected, it should be prepared for by inclusion of unit prices or a cost-plus price for the item in the bid forms or contract, but without forming part of the bid total. An example would be a unit price for permanent casing in a contract in which it is not expected that permanent casing will be required; or unit prices for various ranges of quantity of rock excavation in a site where rock is geologically possible but was not found in the exploration program. Such bid unit prices should be scrutinized by the geotechnical engineer before the contract is awarded. If they are "out of line," they should be renegotiated before the bid is accepted.

Sometimes such unit prices are not included in the bid documents but are written into the contract, usually for the specific purpose of avoiding unbalanced bids. These unit prices should receive as careful consideration by the geotechnical engineer as those in the bid forms; in fact, they would probably be more difficult to renegotiate later, or to revise in court action, than would unbalanced prices originating with the bidder.

7.3.   Caution in Selecting Foundation Type

The sensitivity of the cost of drilled pier foundations to geologic details, weather, and equipment type and condition has been mentioned in several places in the preceding chapters. The claims for "extras" and "changed conditions" that are possible when a drilled pier job "goes sour" can be very large.

In circumstances where prolonged bad weather could interfere substantially with drilled pier construction, it might be necessary to change from piers to a pile foundation in order to keep to construction schedules. If this is a possibility, it should be recognized in advance and provided for as a contingency in the general contract, with pile foundation designs ready for use if necessary and contingent bid prices to match. If a change of this sort has to be made during construction without advanced preparation, it is almost sure to result in substantial increases in cost and time.

The authors have seen several instances where a pier-drilling contractor has approached an owner, sometimes after a contract for another type of foundation has been let, and persuaded him to change from a pile foundation to drilled piers by promising a substantial saving in cost. The geotechnical engineer should be consulted before any such move is consummated, and if the owner goes ahead with a change against his advice, he should withdraw from the job at once. In one instance where the authors' firm stayed on the job after such a change, the contractor valiantly struggled with adverse soil conditions to complete the job, but took so long to do it that the increased costs of engineering construction review and inspection substantially exceeded the contract saving. The reader is invited to guess who felt the weight of the owner's dissatisfaction.

7.4.   Supervision, Inspection, and Construction Costs

The competence and performance of the geotechnical engineer and the foundation inspectors can go far toward minimizing claims of "changed conditions." The geotechnical engineer should be assigned the responsibility for planning and adjusting the exploration program, and for its reporting. He should review plans, specifications, and bid documents as they relate to presentation of subsurface information and to drilled pier operations; and he and the foundation inspectors should keep in mind at all times that their operations and reports are the basis for the owner's, the structural engineer's, and the architect's day-to-day understanding of the contractor's progress in relation to subsurface conditions. During construction the geotechnical engineer should be immediately aware of any deviations—or apparent deviations—of ground conditions from those anticipated from the earlier exploration and postulated in the contract documents; and he should make decisions promptly, without delaying the contractor's operations and, on the other hand, without approving any unnecessary work that the owner will have to pay for.

To this end, both geotechnical engineer and foundation inspector should know the assigned allowable bearing pressures for the various expected strata, and also the basis for these allowances—whether governed by code, local experience, conservative estimates, or field or laboratory test results. Where local geologic variations result in a question of whether or not the stratum reached is the same as—or equivalent to—an assigned bearing layer, a decision must be made whether to stop or to continue drilling until a more competent (or perhaps, a more familiar) formation is reached. This decision should be made by someone who understands both the support requirements of the structure and the structural properties and behavior of soil (or rock) in all variations that might be encountered. This is the geotechnical engineer's duty; only to a limited extent can such a decision be delegated to an inspector, and then only to the extent that possible geologic variations can be pre-identified for him. If this type of decision is left to the contractor, you may be sure that he will not take a chance: he will go deeper "just to be sure" and the owner's budget will suffer, sometimes quite unnecessarily. This is not meant as a criticism of drilling contractors. Their ideal task is to drill a clean, dry hole, of dimensions, location, and depth as directed; and to place in it steel and concrete in such a way as to ensure the continuity of load transmission. If a contractor can do this and keep to the construction schedule, his job is well performed. He should not be asked—or permitted—to do the work of the geotechnical engineer.

Two points that require special care in inspection of drilled pier holes are cited below. Inadequate performance of inspection in either case has resulted in major increases in foundation cost.

One aspect of inspection that requires special care is the decision on where to stop when drilling in soft or weathered rock. The load-supporting capacity of even soft rock is usually much higher than is generally supposed. Current design practice is moving in the direction of taking advantage of this property, with factors of safety against failure due to overload approaching 5, 4, or even 3 instead of 10 or 20 as formerly used (though not recognized). At the same time, pier-drilling machines are developing in power and cutting ability, so that a rock that might have represented "refusal" to the largest machine available locally last year might be readily drillable with today's equipment. The foundation inspector must be prepared to recognize and designate the assigned bearing stratum or bottom depth from inspection of the cuttings *plus descent into the hole* and examination of sides and bottom— *not from difficulty or ease of drilling*. Unless geotechnical engineer and foundation inspector are alert to this aspect, pier holes may be taken to entirely unnecessary depths when more than normally powerful equipment is used.

The precise opposite is true when the drilling equipment is not powerful enough, or the cutting bits or teeth are not suitable: "refusal" may be encountered before assigned depth is reached, and the contractor may have to go to jackhammer work ("dental work") to take the hole down. The cost of rock removal by this means may increase by $1000 per cubic yard (or even more), and a claim for "extras" or "changed conditions" is almost inevitable. This points up the importance of specifying minimum size or power of equipment in certain cases (Chapter 6), and the importance of making at least some of the preliminary exploration borings with large-diameter augers (Chapter 5).

Another typical circumstance requiring special care in inspection is the presence of seams of silt or fine sand in a formation that is nominally a stiff or hard clay. Minor seams of silt below the water table may have an alarming appearance to an inspector, yet be of no significance as far as the load-bearing ability of the formation, either in end-bearing or sidewall shear, is concerned. In geologic circumstances where this condition can occur, inspectors need special instructions regarding its significance with respect to their assignment, and geotechnical engineers need to keep in close touch with the inspection. It is reported that in the Chicago area, where silt seams are common in the Wisconsin and Illinoisan till, there was a period when overruns in drilled pier costs were very common because the foundation inspectors employed were using a pocket-size instrument for testing the shearing strength of the clay down in the hole, and did not recognize that the instruments had no legitimate application to the silt layers that were encountered.

The increased cost of construction that can result from inadequate or improper engineering supervision and inspection is not the only excessive cost that may be incurred. The prime objectives of these services are to obtain piers with good bearing in competent earth or rock and to obtain piers which are composed of competent, continuous concrete from bottom to top. Failure to accomplish either of these objectives can result in structural failure or disastrous reconstruction costs. The magnitude of these risks is illustrated by the following quotation from *Engineering News Record* (September 29, 1966, p. 15):

> At 500 N. Michigan Avenue (Chicago), a 25-story reinforced concrete office-apartment skyscraper, the top segments of the defective caissons are being removed and replaced.
>
> On the 25-story building, cracking in the lower flat-plate concrete floor slabs led to the discovery of a major defect in a 4-ft-dia column caisson. The top of the caisson had settled about 1¾ in. when construction reached the 23rd floor. Core borings in the caisson . . . led to the discovery of a clay-filled void in the concrete of the caisson shaft, about 45 feet below grade.

The example is not cited as an instance of inspection failure. The authors have no knowledge of the cause of the imperfections in these piers.

It is cited to illustrate the possible magnitude of the damage which could conceivably follow when technical supervision and inspection have been inadequate or inexpert—or absent. The sky's the limit!

7.5. When a Drilled Pier Job "Goes Sour"

Trouble on a drilled pier construction job usually takes the form of impeded progress, or failure of the pier contractor to keep to schedule. Possible reasons, as has been indicated above, are many. A few that the authors have encountered:

Prolonged bad weather
Subsurface conditions different from those expected
Foundation type or design unsuited to ground conditions
Inadequate equipment
Inadequate or incompetent inspection
Inadequate experience of contractor's men
"Quick" or flowing silt, resulting in loss of ground
Excessive water inflow (often from gravel or cobble layers)
Unexpected rock or hard, pinnacled, or broken rock

Most of the discussion in the preceding paragraphs of this chapter has been concerned with avoiding delays due to any of these reasons. Of course, these reasons for delay should never be allowed to develop; but they do sometimes develop, and the question then arises: What to do about it?

When prolonged bad weather causes delays on a drilled pier job, the delays in subsequent construction may dictate special equipment or provisions to facilitate drill rig mobility, or a change to another type of foundation. As mentioned earlier, pile-driving rigs can sometimes operate under weather conditions which shut down pier-drilling operations. A change of this sort would be expensive, but might be necessary, for example, to allow the structure to meet occupancy commitments.

Subsurface conditions different from those expected should, ideally, never be encountered, but the subsurface is full of surprises. When such a condition does become evident, there should be a prompt meeting between owner, designer, and supervising engineer; and the geotechnical engineer should be a party to the conference and should bring all his experience and expertise to bear in helping decide what changes should and can be made. If the foundation design is not right for the site, this should be recognized and corrected as quickly as possible—regardless of whose fault it is.

If the contractor's equipment proves inadequate for the job, or if his men are not skilled at their jobs, it may be necessary to bring in another contractor. It should be recognized that the deposed contractor will probably claim "changed conditions"; and this points up the necessity for special care in documenting the new contractor's methods, equipment, and progress, and for very careful logging of some of the pier holes that he drills, for comparison with the information originally available to the first contractor. Ordinarily, detailed logs of pier holes are not made. In this instance, such logs might be of substantial help in litigation.

Excessive water inflow may be very difficult to control, especially when the water is carried in a layer of cobbles or broken rock over sound bedrock. When the sealing off of water becomes too difficult or time consuming, it may be best to let the water rise to equilibrium in this hole and place the concrete by tremie or by pumping. This adds to the cost of the concrete, but may be a more acceptable solution than struggling with the bottom seal until it is finally tight.

Unexpected rock conditions that require hand-tool excavation can be very troublesome. Roller-bit tools, patterned after oil-well-drilling tools but using air instead of water or mud to cool the bit and remove the cuttings, are now available from several manufacturers (see Appendix B).

When a site proves geologically unsuited to drilled piers, as demonstrated by the contractor's first efforts, another type of foundation should be designed and a suitable contract let as quickly as practicable. In such a case there should be careful consideration of the possibility that the contractor's equipment or techniques may be at fault. But a site that is really unsuited to drilled pier construction can produce no end of trouble if efforts to overcome all its difficulties are persisted in. The authors recommend prompt decision and redesign in such a case.

7.6.   The Professional Liability Problem*

Engineers and architects have become increasingly—and painfully—aware of the professional liability problem applicable to their professions. Geotechnical engineers have been particularly affected. Available liability insurance limits have been dwindling, while premium expenses and deductibles have been drastically increasing. It has been pointed out that drilled pier construction is particularly sensitive to claims. Therefore the professional liability problem merits discussion.

7.7.   The Evolution of the Liability Problem

Until about the mid-1950s professional liability suits against engineers and architects were rare. The huge increase in such suits is probably a byproduct

* Most of the following material was developed either jointly or concurrently by the authors and Risk Analysis & Research Corporation, Edward B. Howell, President. We want to thank Risk Analysis & Research Corporation for permission to use the portions developed by them.

of a number of factors. For one thing, since the mid-1950s there has been a tremendous increase in the total amount of construction. This almost automatically implies that a lot of new people, many probably lacking in experience, were becoming involved in the industry. Also, new specialties developed, and new types of engineers and subcontractors were appearing on the scene. The pace of the individual project was stepped up because of new equipment, and also because increasing costs made it imperative to have the project completed and functioning as rapidly as possible.

"Fast-tracked" projects brought a new wave of problems. Once a decision to go ahead with construction is made, owners require that a plant go "on stream" (be in production) so as to meet a very specific time schedule. For example, a brewery has to be able to meet the summer beer demand, a shopping center has to meet a fall opening date, or an automobile assembly plant has to be in operation by the time the new models are being introduced. If such schedules are not met, huge losses result. Under these circumstances there is tremendous pressure to complete the project within the allotted time. Possible delays due to weather, or to ground conditions different from those shown in the plans, or to inept inspection or engineering supervision can be very harmful to the owner under these conditions.

These developments have had a serious effect on communications between all of the parties involved. In the days when there were only one or two engineers, an architect, one or two contractors, and the owner on a project, all the parties knew one another. If something went wrong, it was comparatively easy to get together and work out a solution. With the increased number of people involved and the stepped-up pace, it is no longer possible—or at least it is extremely difficult—for all the people involved to know and properly communicate with one another.

The result of all these developments has been a large increase in the number of suits against engineers and architects. This situation caught engineers, architects (and other professionals), and their insurance companies unprepared. It became apparent that large judgments were possible, and an avalanche of suits followed. The United States also has entered an age of "consumerism," in which people expect every article to be "guaranteed." As a consequence there has been an effort by some attorneys to extend the doctrine of implied guarantees (or warranties) of products to cover services provided by engineers and architects.

The design professions have now learned that it is unwise to take on responsibilities for which they have not contracted and are not getting a fee. Engineers have to be extremely careful not to create "pockets of exposure" in their reports, plans, and specifications—and particularly in their construction review and inspection, when these functions are part of the contract. It is important for the geotechnical engineer recommending or inspecting drilled piers to exercise extreme care. As a first step, it would be prudent for him to insist on following through with the field inspection on any project where he has submitted a soil or foundation report calling for the use of drilled piers. It is also essential to have properly trained foundation inspectors on the job, and to have the geotechnical engineer immediately available at all times. Many problems can be avoided if good communications are maintained with the other parties involved, and if any problem that does occur is attacked immediately.

7.8. Liability to Third Parties

It is well to keep in mind all parties that might be harmed by any error or omission—real or alleged—on the part of the engineer or architect. One usually thinks of the owner, the party with whom the architect, engineer, and contractor have contracted, as the only one to whom they are responsible. Unfortunately, this is far from true. Many buildings, for example, are built by one party and immediately sold to another. The land for which an investigation was made may be sold, and a third party can build on it without the knowledge of the geotechnical engineer who made the report. Yet this third party may place real or alleged reliance on the opinions expressed in the report. Or, in a different situation, a contractor—or a drilling subcontractor—can sue the owner for extra charges and the owner can in turn try to collect from the engineer or architect by means of a cross complaint.

If drilled piers are used as shoring, a claim may be filed by an adjacent property owner, alleging that the material removed during construction of the piers removed lateral support of his building and caused settlement. The architect or structural engineer also could be sued by any of these parties, and he in turn could file a cross complaint against the geotechnical engineer. The insurers for any of the unknown third parties, or lenders financing the construction to which damage allegedly occurred, also can file cross complaints against any of the various parties involved—including the geotechnical engineer. Or a suit may be instituted on behalf of an "injured workman," someone employed by the contractor, whose recourse against his employer is restricted to the limited recovery of workmen's compensation, while his potential recovery against others is unlimited.

7.9. The Cost of Professional Liability Claims

As indicated in the previous paragraphs, suits or claims concerned with construction may—and generally do—involve many parties. For example, an injured third party, when filing a suit for alleged damage caused by drilled pier construction on an adjacent property, could name the owner, the contractor, the drilling subcontractor, the architect, the structural engineer, the geotechnical engineer, and probably others. This multiplicity of defendants makes the defense of such claims extremely expensive for all concerned. Each of the parties listed in the example above may have one or more attorneys representing him. (For example, a defendant may use his corporate attorney to protect the deductible portion of the claim, while another attorney may represent the portion covered by insurance.) As a result 10 or more attorneys may be attending seven or more sets of depositions. A current reporter, together with numerous copies of transcriptions of the testimony given, is involved in each deposition. Many of the parties may hire their own "outside experts." If the case goes to court, because of the number of parties participating in the suit, a lengthy trial often results.

Because of such complications, the cost of defense accounts for the major portion of the liability losses suffered by engineers and architects in all claims, including those involving drilled piers. It is not to be inferred that no major judgments have been rendered against engineers or architects for alleged errors or omissions connected with drilled pier installations or other projects. There have been large judgments rendered, and probably a fairly substantial number of small ones. The principal cost of professional liability claims against engineers and architects, however, has historically been connected with defense.

7.10. Loss Prevention Practices

The best way to avoid a claim, of course, is to have no problems. Drilled pier construction, unfortunately, tends to be somewhat problem prone. It is important then for the geotechnical engineer to keep three significant objectives in mind when working on a project involving drilled piers: (1) to plan and carry out a comprehensive investigation, thoroughly covering all possible problem areas, both in the investigation and in the report; (2) to make certain that the owner is aware of the limitations on the degree of certainty of the findings reported, as well as the effects of weather, adequacy of equipment, and contractor's experience; and (3) to insist on carrying out the responsibilities of construction review and inspection with properly trained personnel who are able to maintain effective communications with the other parties on the job.

Many suggestions for accomplishing the above objectives have been discussed previously. In making the field investigation, geology, as well as the use of geophysical and other remote sensing methods (when applicable), may be helpful. The owner should be made aware of the fact (and it should be clearly spelled out for the benefit of the drilling contractors submitting bids) that a boring log represents only the soils *at the location of the boring, and only at the time the boring was made*. (For example, a water table may vary with time.)

Geotechnical engineering is not an exact science. Even though a most comprehensive investigation may be undertaken, the actual soils sampled can represent only a very small fraction of the total amount of material that will be encountered during actual construction. Natural soil deposits can be extremely variable, and extrapolations between borings should be classified only as "best estimates." The exact conditions are not fully known until the excavations have been made and the foundation completed.

The owner and the engineer or architect preparing the contract plans should be advised not to put the boring logs on the plans. The bidder should be advised as to where he can examine the samples (which should be properly preserved in airtight containers), and he should be furnished with a complete copy of the report describing the complete geotechnical investigation, including the boring logs. Not only can errors or omissions be made in copying logs, but the report is necessary for proper interpretation of the boring logs.

It should be stressed to the owner that the experience and reliability of the drilling subcontractor, together with the adequacy of his equipment, are extremely important. If the job has a strict time schedule, the owner should be made aware of the problems that can be caused by adverse weather, and contingency plans should be developed for emergency use. Training of field technicians—both in technical aspects and in good communications—cannot be overemphasized, nor can the primary necessity of doing careful inspection and construction review.

7.11. Contract Language

The engineer and architect should be aware of the provisions to be included in the contract between the owner and contractor, particularly relating to "changed conditions" or "extras" (discussed previously). The pier-drilling subcontractor should be aware of these provisions as well. It is advisable that language be included in the contract between the owner and the contractor, holding the engineer and architect harmless from injuries or damage arising out of the contractor's operations and not due to the *active* negligence of the engineer or architect (the injured workman claim, for example). To accomplish this protection, it must be made clear to the owner, and the contract language should specify, that any "supervision" or "inspection" performed by the engineer or architect is only for the purpose of obtaining general compliance with the plans and specifications, and that the engineer and architect *do not* and are not expected to have any supervision whatsoever of the contractor's operations. It should be stated in both the A/E-Owner contract and the construction contract that job safety is the sole responsibility of the contractor.

Courts tend to strictly limit the application of "hold harmless" clauses as being against the public interest. Attorneys can assist in drafting clauses which are sufficiently clear to pass the test of strict construction. They can also advise as to the public policies of particular states.

7.12. Limitation of Liability

Limitation of liability for direct damages to the owner arising out of errors or omission by the engineer is possible if included in the contract between engineer and owner. Laws governing limitation of liability, like other laws, vary from state to state. Attorneys should be consulted with respect to local law.

7.13. Statutes of Limitations

Statutes of limitations can be helpful in some cases, but again they vary considerably from state to state. The courts in some states have a tendency, however, to be very liberal in their interpretation of the statutes, particularly with respect to innocent (or presumably innocent) third parties. In some states, for example, the courts have held that the statutes do not start to run against third persons from the time of the completion of the project, but rather from the time of discovery of any latent damage. Furthermore, in the case of continuing damage (e.g., settlement of adjacent structures, which may continue for months or even years), it has been held that the statutes do not start to run as long as damage continues to occur.

As an example, an existing building adjacent to a heavy structure on drilled piers might settle due to the additional load imposed on the underlying strata. The condition may be latent (present but not readily visible) for several years before the damage progresses to the extent that it becomes apparent. It could continue for several more years. The interpretation of the courts in some states may be that the statutes of limitations do not start to run against the owner of the adjacent building until the damage has stopped. In such a case the injured party could file a suit perhaps 15 years after the completion of the structure, naming all parties who conceivably could have contributed to damage. To overcome this problem, many states have enacted special statutes of repose designed to preclude construction claims after a number of years following completion. These statutes vary in scope, conditions, and duration. Again, counsel should be consulted.

7.14. Implied Warranty

An earlier mention was made of the tendency on the part of the public (fostered by decisions reached by courts, consumer suits, and new laws designed to protect the consumer) to expect that everything they purchase is warranted to be satisfactory for the purpose intended. There actually have been attempts to extend this "implied warranty" to persons rendering services, such as engineers. Generally, however, the courts have continued to apply the so-called "prudent man" rule. This requires that an engineer giving recommendations for drilled piers (for example) exercise the same degree of judgment and care that any reasonable and prudent engineer would exercise under the same circumstances—that is, in the same or similar area and at the time that the services were provided. In view of the present tendency to protect the consumer, good judgment would seem to suggest that nonwarranty provisions be included in proposals, reports, specifications, contracts, or any other documents pertaining to an engineering project, to make clear that no warranty is intended.

7.15. Summary

In summary, it is prudent, when drilled piers are called for in a design, for the engineer and architect to exercise careful judgment both in the engineering and in the contract language; to make full disclosure of the degree of reliance that should be placed on the geotechnical information, the possible effects of weather, and the influence of the contractor and his equipment; and to follow through with competent construction review and inspection.

The engineer or architect should not rely on his own knowledge when legal interpretation is involved, but for his own protection should consult with an attorney. This is especially important insofar as protective clauses in contracts are concerned. As previously mentioned, laws governing such clauses (as well as limitation of liability, nonwarranty provisions, and statutes of limitations) vary considerably from one state to another, and the advice of competent attorneys is essential to engineers and architects in protecting themselves against liability claims.

As previously stated, the best way to avoid a claim is to have no problems. Unfortunately, once someone sustains any damage, everyone who conceivably could have contributed to the alleged damage can be named in a suit. A good plaintiff's attorney is required to do this to protect his client. (Naming multiple defendants also is an inexpensive way to gather information by means of depositions.) Once named, the engineer or architect must defend the suit, even though he may have made no contribution whatsoever to the damage. In view of these circumstances, it is not enough for the individual engineer or architect to exercise extreme care; in addition, every attempt should be made to get all the parties involved in the project together at its inception, particularly when drilled piers are involved. All possible problem areas should be discussed, so that everyone concerned with the project has a complete understanding of them. If a friendly attitude and good rapport can be established and maintained, problems that do arise generally can be resolved without the need for a claim to be filed.

Appendix A

Characteristics of Large Drilling Machinery (Manufacturers' Data, 1985)

Maximum hole diameters and depths listed in the following tabulations are manufacturers' estimates for assumed drilling conditions, and torque values given are those provided by the manufacturers. There are no standard specifications for determination of these values, and comparison of one manufacturer's assigned limits with those of another may not be entirely valid. Nonetheless, the authors believe that the following data will be helpful in evaluation of various machines suitable for a particular task. Deeper and larger holes can often be drilled by use of drill stem extensions, special reamers, and so on.

Characteristics of Calweld Drills (1984): Crane-Mounted Rigs to Use Either Drilling Bucket or Auger

Model	Maximum Hole		Maximum Continuous Torque (lb-ft)	Maximum Downward Force (lb)
	Diameter (ft)[a]	Depth (ft)[b]		
200 CH	24	Depends	400,000+	36,000
175 CH	16	on crane	176,500	plus
155 CHS	16	and kelly	175,000	weight of
125 CH	8		125,000	kelly and
80 CH	6		80,000	tools
55 CH	4		50,000	

[a] Maximum hole diameters given are for drilling buckets with reamer.
[b] Maximum depths are for longest kelly (deeper holes can be drilled).

176

Characteristics of Calweld Drills (1984): Truck- and Crawler-Mounted Rigs for Bucket Only

	Maximum Hole	
Model	Diameter (ft)[a]	Depth (ft)
Mechanical		
500 C	11	100
250 C	11	100
200 C	10	85
175 C	8	70
150 C	7	70
All hydraulic		
4500 LH	12	145
52 LH	10	115
45 LH	8	115
42 LH	8	115
45 MH	8	72
42 MH	8	72

[a] Maximum hole diameters are for bucket with reamer.

Characteristics of Calweld Drills (1984): Truck-Mounted Auger Rig for Auger Only

	Maximum Hole	
Model	Diameter (ft)	Depth (ft)
ADL	6	100

Tables courtesy of Calweld, Inc., Santa Fe Springs, CA.

Characteristics of "Williams Diggers"[a]

Model	Mounting	Stall (torque, lb-ft)	Maximum Continuous Force (lb)	Maximum Hole Diameter (ft)	Depth (ft)	Kelly Bar Size (in.)
			Heavy duty			
LLDH	Truck	100,000	50,000	10	120	7/5¼[b]
LDH	Truck	50,000	37,000	8	100	6/4¼
			Medium duty			
DRILL SGT	Truck	37,600	26,000	6	35–60	6/4¼

[a] Table courtesy of Atlantic Equipment Company, Gainesville, VA. (These machines formerly manufactured by Hughes MICON.)

[b] Double figure eleven/seven indicates telescoping kelly, inner shaft 7-in.-square solid steel, outer 11 in. square (outside measurement).

Characteristics of Texoma Holediggers (1984)[a]

Model	Mounting	Maximum Continuous Torque (lb-ft)	Maximum Continuous Down Force (lb)	Maximum Hole		Kelly Bar Size (in.)
				Diameter (ft)	Depth (ft)	
Taurus	Crane carrier	83,740	51,200	8	100	7/5½
700	Truck or crawler	53,990	22,500	6	60	5½/4
600	Truck	39,030	26,000	6	35	3
330	Truck	39,030	26,000	6	20	3
270	Truck	20,200	16,000	4½	20	2½
Economatic	Truck	5,500	35,400	4	20	2½
Heavy-duty economatic	Truck	11,600	35,400	4	25	2½ or 3

[a]Table courtesy of Reedrill, Inc., Sherman, TX.

Specifications and Dimensions for TONE Rodless Reverse Circulation Drill[a]

Model		RRC-15	RRC-20	RRC-30
Drilling Capacity				
Hole diameter	(mm)	1000 1200 1270 1400 1500	1500 1600 1800 2000	2300 2500 2800 3000
	(in.)	39 47 50 55 59	59 63 71 79	91 98 110 118
Depth	(m)		Standard 50, Maximum 80	
	(ft.)		Standard 164, Maximum 260	
Height of motor and drill	(mm)	3675	3675	3900
	(in.)	144.68	144.68	153.54
Bit rotation	(rpm)	32 (50 Hz)	22 (50 Hz)	17 (50 Hz)
I.D. of reverse pipe	(mm)	150	150	200
	(in.)	6	6	8
Power required	(kw)	15 × 2 sets	18.5 × 2 sets	30 × 2 sets
Weight of motor drill (approx.)	(kg)	9,000	12,000	18,000
	(lb)	19,800	26,400	39,600

[a] As of 1982, this machine was available (on contract basis) from Mergentime Corporation, Flemington, NJ.

179

Characteristics of W-J Crane Attachment Caisson Drill (1984)[a]

Model	Maximum Hole Diameter (ft)	Depth (ft)	Maximum Continuous Torque (lb-ft)	Maximum "Crowd" (lb)
SS 8487 single unit	12	200	150,000	[b]
W-J twin 300 unit	18	250	300,000	[b]
W-J super single M-4 unit	26	300	750,000	[b]
W-J super single M-5 unit	35	1,000	1,000,000	[b]

[a] Table courtesy of W-J Sales Co., Inc., Bellville, TX.
[b] Maximum downward force ("crowd") depends on type of mounting and weight of unit and crane.

Characteristics of Watson Drilling Rigs (1984)[a]

Model	Mounting	Torque (lb-ft)	Crowd (lb)	Diameter	Depth	Outer Kelly
6000	Crane attachment	187,000	60,000	12	[b]	12 in.
5000	Crane attachment	98,000	40,000	8	[b]	8 in.
3000-110	Crane carrier or crawler	103,000	50,000	8	110	8 in.
3000-80	Crane carrier or crawler	103,000	50,000	8	85	8 in.
2000-90	Truck or crawler	82,000	30,000	6	90	7 in.
2000-60	Truck or crawler	82,000	30,000	6	60	6 in.
1500	Truck or crawler	64,000	24,000	5	60	6 in.

[a] Table courtesy of Watson Incorporated, Fort Worth, TX.
[b] Depth depends on crane capacity. Approximate depths with practical kelly lengths are 280 ft for Model 6000 and 215 ft for Model 5000.

Appendix B

Information Sources

Following is a list of suppliers and organizations who have furnished information, help, and encouragement for the compilation of this book. We have listed them under particular categories of products or services because we found in many cases that it was difficult to locate sources of information or materials in these fields, and we believe that the inclusion of addresses and telephone numbers may be helpful to many readers. We have not attempted to compile a complete listing for any category, nor have we listed foreign sources for materials or machinery, except in instances where specific experience has led us to believe that a particular machine, material, or practice would be useful in the United States.

Pier-Drilling Machines—Manufacturers

Acker Drill Company, Inc., P.O. Box 830, Scranton, PA 18501, (717) 348-0466.

Atlantic Equipment Company, P.O. Box 186, Gainesville, VA 22065, (703) 754-7114. Williams drilling machines (formerly Hughes MICON); drilling augers, bits, tools.

Bauer-Schrobenhausen, Karl Bauer Spezialtiefbau, GmbH and Co. KG (Ltd.), P.O. Box 1260, D-8898 Schrobenhausen, Federal Republic of Germany. Grouted piles (drilled piers with bottom and side grout injection).

Bay Shore Drilling Systems, Inc., P.O. Box 546, Benicia, CA 94510, (707) 745-9233. Tieback drills.

Calweld Inc., 11212 South Norwalk Blvd., Santa Fe Springs, CA 90670, (213) 863-9377. Large and small bucket and auger rigs; truck and crane mounted; augers, buckets, tools.

Dunham Mfg. Co., Inc., P.O. Box 430, Minden, LA 71055, (318) 377-3535. Small hydraulic auger rigs.

Gus Pech Mfg. Co., Inc., 1480 Lincoln St. SW, Le Mars, IA 51031, (712) 546-4145. Small truck-mounted bucket rigs.

Hughes MICON, 3001 South Highway 287, Corsicana, TX 75110, (124) 872-5671. Hughes MICON Combination Shaft Drills (for very large holes).

Reedrill, Inc., P.O. Box 998, Sherman, TX 75090, (214) 786-2981. Texoma Holediggers (truck mounted).

Strata-Dyne, 2704 N. Nichols, Fort Worth, TX 76016, (817) 625-1531. Strata Drills (truck and crane mounted).

Tone Boring Co. Ltd., 6, Meguro 1-chome, Meguro-ku, Tokyo 153, Japan, Tokyo 493-0111. Rodless Reverse-Circulation Drill.

Watson Incorporated, P.O. Box 11006, Forth Worth, TX 76109, (817) 927-8486. Large and small rigs, truck, carrier, crane, crawler mounted; tools and equipment.

W-J Sales Company, Inc., P.O. Box 774, Bellville, TX 77418, (713) 497-6793. Large crane-mounted rigs; drilling tools.

Augers, Buckets, Tools, and Accessories—Manufacturers

Drilco Industrial, P.O. Box 3135, Midland, TX 79702, (915) 683-5431. Smith Big Hole Drilling Tools.

Fansteel VR/Wesson, P.O. Box 11399, Lexington, KY 40575, (606) 252-1431. Carbide bits and block units.

H and T Auger Co., P.O. Box 6048, Odessa, TX 79762, (915) 362-4471. Augers, bits, drilling tools.

Ingersoll-Rand, 3 Century Drive, Parsippany, NJ 07054, (201) 267-0476. Down-hole percussion/rotary bits for hard rock.

Steven M. Hain Co., 713 E. Walnut, Garland, TX 75040, (214) 272-6461. Buckets and belling tools.

W F J Drilling Tools, Inc., P.O. Box 6724, Odessa, TX 79762, (915) 366-2514. Augers, core barrels, drilling tools.

Expansive Cement

Con-Rok (International), Kapiland 100, Suite 408, West Vancouver, British Columbia, Canada V7T 1A2, (604) 926-7725. S-MITE expansive cement (nonexplosive demolition agent); manufactured by Sumimoto Cement Company, Ltd., Japan.

Construction and Industrial Supply Co., 189 Cobb Parkway North, Building B Suite 3, Marietta, GA 30062, (404) 429-0008. BRISTAR expansive cement (nonexplosive demolition agent); manufactured by Onoda Cement Company, Japan.

Carriers

Pettibone Texas Corp., P.O. Box 4249, Forth Worth, TX 76106, (817) 232-1050. Carriers for drill rigs, cranes, and so on.

Hydraulic Rock Splitters

ELCO International, Inc., 111 Van Riper Avenue, Elmwood Park, NJ 07407, (201) 797-4644.

Vibro Driver/Extractors

L. B. Foster Company, 415 Holiday Drive, Pittsburgh, PA 15220, (412) 928-3400. Foster Vibro Drivers/Extractors.

International Construction Equipment, Inc., 301 Warehouse Drive, Matthews, NC 28105, 1-800-438-9281. ICE hydraulically powered vibratory driver/extractors.

Pileco, Inc., P.O. Box 16099, Houston, TX 77222, (713) 691-3638. Hydraulic and electric PTC Vibrodrivers.

Desander for Cleaning Slurry

Pileco, Inc., P.O. Box 16099, Houston, TX 77222, (713) 681-3638. Caviem Desander.

Drilling Muds and Slurry Testing Instruments

N. L. Baroid Co., P.O. Box 1675, Houston, TX 77251, (713) 527-1100. A wide variety of drilling muds and mud testing instruments.

American Colloid Company, 5100 Suffield Ct., Skokie, IL 60077, (312) 966-5720. A wide variety of drilling muds, chemicals, and mud testing instruments.

Johnson Division UOP, P.O. Box 43118, St. Paul, MN 55164, (612)) 636-3900. Johnson Revert, a self-destroying drilling fluid additive.

Soil and Groundwater Testing Equipment

Soiltest, Inc. (Environmental Division), P.O. Box 931, Evanston, IL 60204, 1-800-323-1242. A wide variety of soil and groundwater testing equipment and instruments.

Static Cone Penetration Equipment

United States Distributor	Manufacturer
Technotest USA, Inc. 33977 Chardon Road, Willoughby Hills, OH 44094	*Technotest, s.n.c.* Via E. de Nicola, 31 41100 Modena, Italy
Hogentogler and Company, Inc. PO Box 385, Gaithersburg, MD 20876	*Goudschinefabriek, B.V.* PO Box 125, Gouda, Holland
Terrametrics, Inc. 16027 West 5th Avenue Golden, CO 80401	*A. P. van den Berg, B.V.* Holland

Depth Measurement Equipment

D.M.T. Inc., P.O. Box 50832, Denton, TX 76206, (817) 566-2146. DD-100 Depth Display Unit.

Organizations

ADSC—International Association of Foundation Drilling Contractors, Box 280379, Dallas, TX 75228.

ASTM—American Society for Testing and Materials, 1916 Race Street, Philadelphia, PA 19103.

ASFE—Association of Soil and Foundation Engineers, 8811 Colesville Road, Suite G106, Silver Spring, MD 20910.

Deep Foundations Institute, P.O. Box 359, Springfield, NJ 07081.

In addition to the suppliers listed in this Appendix there are many manufacturers and supplies of drilling rigs, tools, and materials to the foundation drilling industry. The best compilation known to the authors is the directory of the ADSC, which has four classes of membership:

1. Contractor members.
2. Associate members—suppliers to the Foundation Drilling Industry.
3. Technical affiliate members—engineers, designers, technical services.
4. Honorary technical affiliate members.

Appendix C

ADSC Guidelines for Specification and Inspection

and

Standards and Specifications (1983) for the Foundation Drilling Industry*

Industry Guidelines

Preamble

This paper represents the opinions of a joint committee consisting of design engineers, inspection agencies, and contractors. The following pages represent a coordinated view of drilled shaft construction which addresses the technical concerns of the engineering community designing drilled shafts while recognizing the practical limitations of existing technology, equipment, and technique. It should be recognized that the imposition of specifications which are not economically feasible or technologically possible given the existing state of the art does not increase quality, and serves only to discredit the capabilities of a proven foundation system—the drilled shaft.

Inspection

The quality of a drilled shaft is governed by its installation. A proper choice of installation procedure and equipment, good workmanship, and tight control of all installation work are essential to the construction of a good drilled shaft.

---

* Reprinted with permission of ADSC (International Association of Foundation Drilling Contractors).

186

Purpose

Inspection shall be carried out and paid for by the owner or by another suitably qualified person responsible to the owner to ensure that subsurface conditions are consistent with the design and that construction is carried out in accordance with the design and with good engineering practice.

The Inspector

It is essential that inspection personnel be well experienced in this field so as to be able to verify adherence to plans and specifications and to properly evaluate actual soil conditions in drilled shafts, while recognizing that work should be carried out in a timely manner.

Preparation

Good inspection begins prior to actual construction with the examination of all design documents. The following should be available to the inspector on the site:

Soil Investigation Report (Subsurface Investigation Report, Geotechnical Report)
Drawings of the foundations
Specifications
Basis of measurement and payment
Any other documents on special design features or assumptions
Loading information

Inspection for Dry Construction Method

The contractor shall provide equipment for checking the bearing, dimensions, and alignment of each drilled shaft. The dimensions and alignment of the shaft shall be determined by the Contractor under the observation of the Inspector. The bearing is determined by the Inspector with the assistance of the Contractor. Generally, these functions may be checked by any of the following methods:

1. Check the dimensions and alignment of the shaft excavation using reference stakes and plumb bob to determine the center of the shaft.
2. Check the dimensions and alignment of the bottom casings, if and when they are inserted in the excavation.
3. Other means provided by the Contractor and approved by the Owner's representative.

The following tolerances of the excavation are recommended:

1. Vertical tolerance of 2% plumbness over the entire length of the completed shaft.
2. The top of the shaft within 3 in. of the required location.

Bearing stratum for end-bearing shafts is frequently determined by one or a combination of the following:

1. The Subsurface Investigation Report (Soil Investigation Report, Geotechnical Report).
2. Shear vane or penetrometer tests in situ.
3. Geophysical measurements.
4. In shafts bearing on rock and if required by the specifications a 1½- to 2-in. probe hole is to be drilled in a representative number of shafts. Depth of the probe hole is normally one to one and one-half (1–1½) times rock socket design diameter but not to exceed ten (10) feet.

How Clean Is Clean?

The amount of loose material left in the bottom of the shaft should be a function of whether the shaft is end-bearing, friction, or a combination of both friction and end bearing.

Cleaning should be required only to the extent that no settlement of the concrete shaft would occur. In areas where hand cleaning is necessary, a man with a shovel and bucket can accomplish this task. Sweeping, mopping, sponging, and washing the bottom of the shaft are not necessary and should not be required.

Both mechanical cleaning and inspection from the ground surface are recommended when possible. This produces a safer work environment and is economical.

Inspection Methods for Construction of Shafts Under Water or Slurry

The contractor shall provide equipment for checking the bearing, dimensions, and alignment of each drilled shaft. The bearing is determined by the Inspector with the assistance of the Contractor. Generally, these functions may be checked by any of the following methods:

1. Check the dimensions and alignment of the shaft excavation using reference stakes and plumb bob to determine the center of the shaft at ground level.
2. Check the dimensions and alignment of the bottom casings if and when they are inserted in the excavation.
3. Check the dimensions of the drilling tools being used to advance the shaft.

4. Probe shaft with kelly bar or weighted tape.

5. A television camera sealed inside a watertight jacket to inspect the bottom and sides of the shaft. Unless absolutely required this approach should be discouraged, since it adds considerable costs to the job, interferes with the production, and has many technical problems, for example, need of coagulants, and so on to reduce turbidity.

6. Other means provided by the contractor and approved by the owner's representative.

The following tolerances of the excavation are recommended:

1. Vertical tolerance of 2% plumbness over entire length of the excavated shaft.

2. The top of the shaft within 3 in. of the required location.

Bearing stratum is normally determined by one or a combination of the following:

1. Observing penetration into the bearing stratum.

2. The Subsurface Investigation Report (Soil Investigation Report, Geotechnical Report).

Cleaning the excavation is accomplished by one or more of the following:

1. Cleanout bucket.

2. Bailing bucket.

3. Water-jet and/or air-lift pump.

4. Other means provided by the contractor and approved by the owner's representative.

Concrete Placement

Concrete for the shaft should be of an accepted and proven mix design with a 7- to 9-in. slump. Free fall is not an acceptable method of concrete placement in shafts to be concreted under slurry or water. Concrete under slurry or water shall be pumped or placed with a tremie. The tremie for depositing concrete in the drilled shaft shall consist of a tube having a minimum inside diameter of 10 in. or a pump pipe of 4-in. minimum diameter and shall be constructed so that all sections have watertight joints. The discharge end of the tremie or pump pipe shall be constructed so that it is watertight against intrusion from the outside and so that it will readily discharge concrete. The tremie or pump pipe should be supported so that it can be raised to permit free discharge of concrete and lowered rapidly when necessary to choke off or retard the flow. The tremie pipe should be kept full of concrete until work is completed. Vibrating of the concrete is not required. The top of concrete placed should be $+3$ in. to -3 in. if the cutoff elevation is at ground level. Concrete test cylinders should be taken in accordance with the specifications.

Materials

Reinforcing Steel (if required)

Reinforcing steel shall include all bars, spirals, and/or ties, spaced as shown on drawings. Bars shall conform to ASTM specifications.

Cages should be braced to retain their configuration throughout the placing of concrete and the extraction of the casing from the shaft. Job limitations may require some reinforcing to be assembled in place.

During casing removal, movement of cage upward in itself is not a problem but may be symptomatic of a larger problem, that is, arching of concrete in casing and subsequent "cutting" of shaft. Upward movement of concrete should be monitored.

Three-inch minimum coverage of reinforcing with concrete is required to avoid interference with casing removal. Numerous methods are available to accomplish this procedure such as concrete spacer blocks, concrete wheels, steel rebars, and so on.

Concrete

Concrete for a dry shaft should be an accepted and proven design; in an unreinforced shaft it may have a 4- to 6-in. slump; if there is a rebar cage, a 7- to 9-in. slump is appropriate. Concrete should be placed by "free-fall" methods, that is, without the use of a tremie, provided a hopper or other device is used to prevent the concrete from striking the sides of the excavation or reinforcing cage. Concrete slump for use in casings should be 6 to 9 in. If temporary steel liners are required, they may be withdrawn as the concrete is being placed, maintaining a sufficient head of concrete within the steel liner to prevent reduction in the diameter of the drilled shaft due to earth or hydrostatic pressure on the fresh concrete and to prevent extraneous material from falling in from the sides and mixing with the concrete.

Vibrate or "rod" only the top ten (10) feet of concrete. Concrete of 6-in. slump or higher should never be vibrated. Should it become necessary to stop pouring of the concrete in a shaft, the poured concrete should be brought to level surface and dowels inserted into the concrete as directed by the inspector. All laitance should be removed prior to continued placement of concrete.

Alternate Installation

In the event that inflow of water into a shaft cannot be properly controlled, the water shall be allowed to rise in the shaft to balance the hydrostatic head. The concrete shall then be pumped or placed with a watertight tremie as described in the wet construction method.

Typical Caisson Log Information

Once the shaft is completed, the Inspector should complete a log for each shaft showing the applicable items as follows:

1. Date and time excavation started.
2. Shaft location and identification.
3. Shaft diameter as per plans.
4. Shaft diameter as constructed.
5. Ground elevation.
6. Top of concrete as per plans.
7. Top of concrete as constructed.
8. Top of rock elevation as encountered.
9. Top of rock elevation as constructed.
10. Bottom of shaft elevation as per plans.
11. Bottom of shaft elevation as constructed.
12. Location and extent of cavities.
13. Lineal footage of earth drilling and lineal footage of rock drilling.
14. Comments on water condition (i.e., flow volume, hydrostatic head, where encountered).
15. Bearing surface condition and dimensions.
16. Description of test holes.
17. Volume of concrete placed together with concrete supplier and concrete invoice or batch numbers.
18. Date and time excavation completed.
19. Date and time concrete placed.
20. Other comments as deemed necessary for project including any nonstandard methods of construction which may have been required and affect the shaft configuration or construction.
21. Details of any obstruction encountered during drilling.

Note: Any discrepancies should be resolved immediately, especially those items pertaining to pay quantities. Completed reports should be signed by the Inspector, General Contractor, and Foundation Contractor. Copies must be distributed promptly to all of the above parties.

Standards and Specifications

Foreword

The ADSC, after thorough investigation and study, has adopted these standards and specifications pertaining to foundation work performed by members of the Association. Revised May 1, 1980, these changes reflect technical improvement in the industry to insure better products.

The standards, which include classification of drilled foundations, classification of rock excavation for pay purposes, standard sizes of casing and underreams, and samples of on-site paperwork, are the accepted practice of Association members throughout the industry.

The specifications are the recently published ACI (American Concrete Institute) "Standard Specification for End Bearing Drilled Piers," with addenda by ADSC in boldface type. These specifications are equally adaptable to side shear design.

You will notice, however, that ACI refers to a drilled pier in lieu of a drilled foundation. We accept the terms as interchangeable and thank the ACI for their fine efforts in having written such a workable specification.

In the past the absence of standards and specifications has resulted in confusion, misunderstandings, and wide variation in specifying, bidding, and performing drilled foundation work. In many instances costs were higher than necessary because of special requirements which can now be eliminated since these standards are available. The ADSC believes that the specifications and contract for a drilled foundation project should give the contractor as much leeway as practicable in the choice of equipment and methods. A performance type of specification, imposing no unnecessary restrictions but requiring an acceptable end result, will in the long run result in more competitive bids, lower bid prices, fewer disputes and claims for extras, and usually better foundations than would a tight specification that unnecessarily restricts the choice of contractors or impedes the work of the successful bidder.

The ADSC constantly strives to upgrade and perfect procedures which will offer the construction industry a superior and more economical product.

ADSC Standards (revised May 1, 1980; updated August, 1983)

Item 1. Classification for Drilled Foundations

Type A Not cased. Not reinforced.

Type B Not cased. Reinforced.

Type C Temporary tool casing. Not reinforced.

Type D Temporary tool casing with permanent liner. Not reinforced.

Type E Temporary tool casing. Reinforced.

Type F Temporary tool casing with permanent liner. Reinforced.

Type G Permanent casing. Not reinforced.

Type H Permanent casing. Reinforced.

Underreamed Shafts—Add the letter U to the classification type of shaft.

Battered Shafts—Add the letter B to the classification type of shaft.

Underwater Concrete Placement—Add the letter W to the classification type of shaft.

Note: All types include concrete as specified for project. Safety regulations will be followed.

Item 2. Scheduled Diameters for Foundation Drilling

Types A—B		Types C—D—E—F—G—H		
Shaft Size Nominal Diameter	Diameter of Drilled Hole	Outside Diameter of Casing	Diameter of Drilled Hole for Casing	Diameter of Drilled Hole Extended thru Casing
		Minimum Recommendation		
12"	12"	Not Recommended		
14"	14"	Not Recommended		
16"	16"	Not Recommended		
18"	18"	18"	20"	16"
20"	20"	20"	22"	18"
24"	24"	24"	26"	22"
30"	30"	30"	32"	28"
36"	36"	36"	38"	34"
42"	42"	42"	44"	40"
48"	48"	48"	50"	46"
54"	54"	54"	56"	52"

(Table continues on p. 194.)

Shaft Size Nominal Diameter 12"	Types A—B Diameter of Drilled Hole 12"	Types C—D—E—F—G—H		
		Outside Diameter of Casing	Diameter of Drilled Hole for Casing	Diameter of Drilled Hole Extended thru Casing
		Minimum Recommendation		
60"	60"	60"	62"	58"
66"	66"	66"	68"	64"
72"	72"	72"	74"	70"
78"	78"	78"	80"	76"
84"	84"	84"	86"	82"
90"	90"	90"	92"	88"
96"	96"	96"	98"	94"

Note: Geological conditions in some areas may dictate sizes other than those indicated in the following schedule. Consult your local ADSC Contractor for recommendations.

Item 3. Underreamed Shafts

A. Maximum base diameter of an underreamed shaft is recommended, for economical reasons and practicality, not to exceed three times the diameter of the shaft for a fully developed underream.

B. The fully extended underreamed angle recommended is 45 degrees when design and conditions permit.

C. The standard toe height for the underream is 3" for drilled shaft sizes 18" through 42" and 6" for sizes 48" and larger.

Item 4. Temporary Tool Casings for Drilled Foundations

A. Casings are regular grade steel, produced by electric seam welding, butt welding, or spiral welding

B. The outside diameter of casing for each size of drilled foundations is as listed in item 2.

C. Available wall thickness for each standard casing is as follows:

Standard Tool Casings Outside Diameter	Standard Available Wall Thickness Range
18" thru 24"	Min. ¼"; ⁹⁄₃₂"; ⁵⁄₁₆"; ⅜" Max.
30" thru 36"	Min. ⁵⁄₁₆"; ⅜"; ⁷⁄₁₆" Max.
42" thru 60"	Min. ⅜"; ⁷⁄₁₆"; ½" Max.
66" thru 96"	Min. ¹³⁄₃₂"; ⁷⁄₁₆"; ⁹⁄₁₆"; ¾" Max.

D. Tolerances on the outside diameter and other dimensions of casings are standard API tolerances applicable to regular steel line pipe.

Item 5. Qualifications

A. Work must be accomplished by companies employing personnel experienced in drilled foundation work.

B. Experience must be relevant to anticipated subsurface materials, water conditions, shaft sizes, and special technique required.

C. Demonstrate to the satisfaction of the Owner's Representative the dependability of equipment and techniques to be used.

Item 6. Subsurface Investigation

All of the subsurface information available to owner, architect, structural engineer, or geotechnical engineer shall be made available to prospective bidders; and this shall include soil and rock samples from the subsurface exploration, preserved at natural moisture content and arranged so that they can be readily examined. The samples, of course, will have to be kept at some central location, such as the architect's office. It is recommended that the owner employ and pay for all geotechnical services required. A conflict of interest could occur if these services are provided and paid for by the contractor. The geotechnical report (or soil and foundation report), with complete boring logs, should be included in the plans and the specifications and considered a part of the contract documents.

ILLUSTRATION OF MEASUREMENT FOR CLASSIFIED ROCK

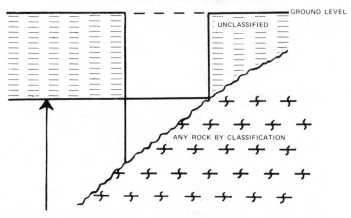

THIS ELEVATION IS "TOP OF ROCK" & SHAFT IS CONSIDERED "ROCK EXCAVATION" FULL VOLUME BELOW THIS ELEVATION:

(*a*)

Classification of Rock for Pay Purposes: Rock is defined as any material which cannot be drilled with a conventional earth auger and/or underreaming tool, and requires the use of special rock augers, core barrels, air tools, blasting, and/or other methods of hand excavation. All earth seams, rock fragments, and voids included in the rock excavation area will be considered rock for the full volume of the shaft from the initial contact with rock for pay purposes. See figures (*a*), (b) and (c).

(*Figure continues on p. 196.*)

ILLUSTRATION OF A DRILLED FOUNDATION EXCAVATION

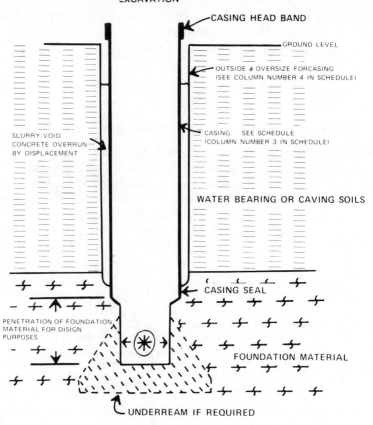

CASING HEAD BAND

GROUND LEVEL

OUTSIDE ⌀ OVERSIZE FORCASING
(SEE COLUMN NUMBER 4 IN SCHEDULE)

SLURRY/VOID
CONCRETE OVERRUN
BY DISPLACEMENT

CASING   SEE SCHEDULE
(COLUMN NUMBER 3 IN SCHEDULE)

WATER BEARING OR CAVING SOILS

CASING SEAL

PENETRATION OF FOUNDATION
MATERIAL FOR DISIGN
PURPOSES

FOUNDATION MATERIAL

UNDERREAM IF REQUIRED

✳ SEE SCHEDULE FOR DIAMETER OF
PENETRATION EXTENDED THRU CASING
(COLUMN NUMBER 5 OF SCHEDULE)

(b)

196

ILLUSTRATION OF A DRILLED FOUNDATION

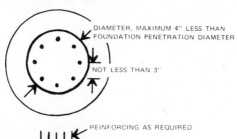

DIAMETER, MAXIMUM 4" LESS THAN FOUNDATION PENETRATION DIAMETER

NOT LESS THAN 3"

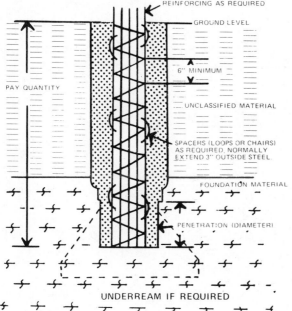

REINFORCING AS REQUIRED

GROUND LEVEL

6" MINIMUM

PAY QUANTITY

UNCLASSIFIED MATERIAL

SPACERS (LOOPS OR CHAIRS) AS REQUIRED, NORMALLY EXTEND 3" OUTSIDE STEEL.

FOUNDATION MATERIAL

PENETRATION (DIAMETER)

UNDERREAM IF REQUIRED

(c)

DAILY REPORT FORM

DATE _____ GENERAL CONTRACTOR _____ JOB No. _____ LOCATION _____

Steel furnished by: _____ Driller _____

_____ G.C. _____

Steel tied by: _____ Driller _____

_____ G.C. _____

Concrete furnished by: _____ Driller _____

_____ G.C. _____

Shaft	Bent No.	Shaft Size	Underream Size	Bottom Elev.	Top Elev.	Pay Depth	Steel Length	Casings Yes/No	Concrete Qty.	Other

REMARKS:

JOB FOREMAN

G.C. REP / INSPECTOR

Single Shaft Record Report

ABC Contractors
123 Elm St.
Anytown, U.S.A
Phone: 234-5678

PROJECT_____   DAY _____

JOB NO. _____   DATE _____

DESIGN

SHAFT LOCATION   _____
TYPE/DIAMETER   _____
TOP ELEVATION   _____
BOTTOM ELEVATION   _____
BELL DIAMETER   _____
LENGTH   _____

ASBUILT

DATE STARTED   _____
     COMPLETED   _____
DIAMETER   _____
TOP ELEVATION   _____
BOTTOM ELEVATION   _____
BELL DIAMETER   _____
ADD OR DEDUCT   _____
TOTAL LENGTH   _____
ACTUAL CONCRETE   _____

REMARKS

DRAWINGS (INCLUDE ELEVS, SIZE, SHAPE, CASING, & ALL OTHER PERTINENT INFORMATION)

APPROVED BY =   **OWNERS REPRESENTATIVE**   _____

            GENERAL CONTRACTOR   _____

            ABC CONTRACTORS, INC.   _____

            OTHER   _____

Standard Specification for End Bearing Drilled Piers by ACI Committee 336 (with additions by ADSC)

Specifications

This specification* covers requirements for end bearing drilled pier construction. It includes delivery, handling, and storage of the casing, excavation, soil testing, placing of concrete, and inspection.

Keywords: bearing capacity, concrete construction, excavation, foundation inspection, piers, placing, quality control, reinforced concrete, reinforcing steels, safety, soil mechanics, specifications, standards, tests, tolerances (mechanics).

Foreword

This foreword is included for explanatory purposes only: it does not form a part of Standard Specification ACI 336.1.

Standard Specification ACI 336.1 is a reference standard which the Architect/Engineer may cite in the project specification(s) for any building project, together with supplementary requirements for the specific project.

This specification is written in the section and three-part format of the Construction Specifications Institute, but with the numbering system modified to ACI requirements. The language is generally imperative and terse.

A specification checklist is included as a preface to, but not forming a part of, Standard Specification ACI 336.1. The purpose of this checklist is to assist the Architect/Engineer's designer(s) and specifier(s) to properly choose and specify the necessary supplementary requirements for the project specification(s).

Preface to Specification Checklist

P1—Standard Specification ACI 336.1 is intended to be used essentially in its entirety, by citation in the project specification, to cover all usual requirements for end bearing drilled pier construction. Individual sections, parts, or articles should not be copied into project specifications since taking them out of context may change their meanings.

P2—Building codes set minimum requirements necessary to protect the public. Some of the requirements in this Standard Specification ACI 336.1 may be higher than some minimum building code requirements to assure the owner the level of quality and performance which he expects the structure to provide. However, adjustments to the needs of a particular project shall be made by the Architect/Engineer's designers and specifiers by reviewing each of the items indicated in this specification checklist and

---

* Reprinted with permission of ADSC (International Association of Foundation Drilling Contractors).

then including their decisions on each as mandatory requirements in the project specification.

P3—These mandatory requirements shall designate specific qualities, procedures, material, and performance criteria for which alternatives are permitted or for which provision is not made in Standard Specification ACI 336.1 if required.

P4—A statement such as the following will serve to make the Standard Specification ACI 336.1 an official part of the contract requirements: End bearing drilled pier construction shall conform to all requirements of "Standard Specification for End Bearing Drilled Piers (ACI 336.1-79)," published by the American Concrete Institute, Detroit, Mich., except as modified by the requirements of this project specification.

P5—The specification checklist that follows is addressed to each item of ACI 336.1 that requires the designer/specifier to make a choice where alternatives are indicated, or to add provisions where they are not indicated, or to take exceptions to ACI 336.1. The checklist consists of one column identifying sections, parts, and articles of ACI 336.1 and a second column of notes to the designer/specifier to indicate the action required of them.

Specification Checklist

Section/Part/Article of ACI 336.1	Notes to the Designer/Specifier
Section 1—General Requirements	
PART 1.1. Scope	Indicate specific scope.
PART 1.2. Definitions	Review; take exceptions, or add to, as required.
PART 1.4. Reference standards	Review applicability of cited references and take exception if required. For Recommended Practice, specify as mandatory if applicable; if not applicable, take exception.
PART 1.5. Project conditions	Review; take exceptions, or add to, as required. Especially note if a pre-job conference is required.
Pricing	Pricing is not a part of the standard specification. But pricing forms of contract documents should provide for contract price based on drilled piers to dimensions and elevations shown. Changes in contract price due to additions to, or deductions from, work shown shall be on basis of unit prices in proposal.

(Table continues on p. 202.)

Specification Checklist (*Continued*)

Section/Part/Article of ACI 336.1	Notes to the Designer/Specifier

Section 2—Materials and Construction

PART 2.1—GENERAL

2.1.2. Submittals

To whom sent, for both Geotchnical Engineer and Contractor?

2.1.3.1. Geotechnical engineer testing

Specify extent and type of testing and inspection.

2.1.3.3. Testing agency

What is to be inspected, and extent and type?

2.1.4. Tolerances

Specify when greater tolerances are permitted, if any.

PART 2.2—MATERIALS

2.2.1. Steel casing

Specify ASTM type and grade to use. State any exceptions to welding of teeth. Show on drawings, size, wall thickness, type, and design of permanent casing.

2.2.2. Reinforcing steel

Specify ASTM type and grade to use.

2.2.3. Concrete

Specify 28-day strength, additives (air entrainment, water reducing), slump limits, curing methods for tops of piers.

PART 2.3—CONSTRUCTION

2.3.1.1. Excavation

Specify any special excavation procedures to be followed, such as use of drilling mud or temporary casing.

2.3.1.2. A—Explore bearing stratum

Specify number or minimum percent of piers to be probed (applicable only when no danger of water blow-in from below through probe hole).

2.3.1.2. B–Inspection and testing

Specify inspection and testing procedures to be followed. (It is recognized that procedures vary in different parts of country depending on prevailing geology and experience.)

2.3.1.2. C—Bells

Show bells on drawings.

2.3.1.5. Loose material

Specify limiting amount of loose material or water permitted in hole at time of concrete placement. Specify any other acceptability requirements.

2.3.1.6. Disposal of excavated material

Specify where.

2.3.2.1. Void space

Specify whether grouting is required of any annular void space outside of permanent casing.

2.3.2.2. Removal of casing

Specify removable or permanent casing.

Section/Part/Article of ACI 336.1	Notes to the Designer/Specifier

PART 2.3—CONSTRUCTION (*cont'd*)

2.3.4.1.	Dewatering	Specify specific dewatering criteria.
2.3.4.2.	Approval to place concrete	Emphasize.
2.3.4.4.	Free fall	Specify any special concrete placement procedures required.
2.3.4.9.	Tremie concrete	Specify specific tremie procedures, such as: minimum 7- to 9- in. slump, maximum ⅝-in. aggregate, continuous tremie pipe, minimum pipe embedment in concrete at all times, continuous concrete placement, and static water level in hole prior to concrete placement.
2.3.4.10.	Concrete tests	Specify requirements for making test cylinders and for testing.

Section 1–General Requirements

1.1. Scope

　1.1.1.　This standard specification covers requirements for end bearing drilled pier construction.

　1.1.2.　The provisions of this standard specification shall govern unless otherwise specified in the contract documents. In case of conflicting requirements, the contract documents shall govern.

1.2. Definitions

　　The following definitions cover the meanings of certain words and terms as used in this standard specification.

　1.2.1.　*Acceptable or accepted*—Acceptable or accepted by the Architect/Engineer or Geotechnical Engineer.

　1.2.2.　*Allowable service load bearing pressure*—The vertical pressure per unit area that may be applied to the bearing stratum at the level of the pier bottom. Allowable service load bearing pressure is normally selected by the Geotechnical Engineer on the basis of samples, tests, and applied soil mechanics, with due regard for the character of the loads to be applied and the settlements that can be tolerated.

　1.2.3.　*Architect/Engineer*—The authority, such as the architect, the engineer, the architectural firm, the engineering firm, the contracting officer, or other agent of the owner issuing project specifications and drawings, and/or authorized by the owner to administer work under the project documents.

1.2.4. *Bearing stratum*—The formations or layers of soil or rock that support the pier and the loads imposed on it.

1.2.5. *Bell*—An enlargement at the bottom of the shaft for the purpose of spreading the load over a larger area.

1.2.6. *Casing*—Protective steel casing usually of cylindrical shape, lowered into the excavated hole to protect workmen and inspectors entering the shaft from collapse or cave-in of the sidewalls and for the purpose of excluding soil and water from the excavation.

1.2.7. *Contract documents*—Consist of the agreement, conditions of the contract, contract specifications, contract drawings, and all addenda thereto issued prior to the signing of the contract.

1.2.8. *Contract drawings*—Drawings which accompany contract specifications and complete the descriptive information for drilled pier construction work required or referred to in the contract specifications.

1.2.9. *Contractor*—The organization contracted with to carry out the work shown on the contract drawings and specifications.

1.2.10. *Contract specifications*—The specifications which employ ACI 336.1 by reference and which serve as the instrument for making the mandatory and optional selections available under the specification.

1.2.11. *End bearing drilled pier*—Cast-in-place foundation element with or without enlarged bearing area extending downward through weaker soils or water to a rock or soil stratum capable of supporting the loads imposed on or within it. A shaft diameter of $2\frac{1}{2}$ ft. (0.76 m) is the lower limit for piers covered by these specifications.

1.2.12. *Geotechnical engineer*—The specialized engineer retained by the owner reporting to the Architect/Engineer and with responsibilities as defined herein.

1.2.13. *Testing laboratory*—The testing agency retained by the owner to perform required tests on the contract construction materials to verify conformance with specifications.

1.2.14. *Owner*—Party that pays for approved work performed in accordance with drawings and contract specifications and receives the completed work.

1.2.15. *Permitted*—Permitted by the Architect/Engineer.

1.2.16. *Qualified*—Qualified by training and by experience on comparable projects.

1.2.17. *Required*—Required by the contract documents.

1.2.18. *Shaft*—Drilled pier above bearing surface exclusive of bell, if any.

1.2.19. *Specified*—Defined in the contract documents.

1.2.20. *Submitted*—Submitted to the Architect/Engineer for review.

1.3. Notation

The following abbreviations are defined for use in this standard specification.

1.3.1. *ACI*:   American Concrete Institute
P.O. Box 19150
Detroit, MI 48219

1.3.2. *ASTM*: American Society for Testing
and Materials
1916 Race Street
Philadelphia, PA 19103

1.3.3. *AWS*:   American Welding Society
2501 N.W. 7th Street
Miami, FL 33125

1.3.4. **ADSC: International Association
of Foundation Drilling Contractors
P.O. Box 280379
Dallas, TX 75228**

1.4. Reference Standards

1.4.1.   The standards referred to in this Standard Specification ACI 336.1 are listed in Articles 1.4.2 through 1.4.4 of this Section, with their complete designation and title including the year of adoption or revision and are declared to be a part of this Standard Specification ACI 336.1 the same as if fully set forth herein, unless otherwise indicated in the contract documents.

1.4.2.   *ASTM standards*

A 36-75   Standard Specification for Structural Steel

A 82-76   Standard Specification for Cold Drawn Steel Wire for Concrete Reinforcement

A 252-75   Standard Specification for Welded and Seamless Steel Pipe Piles

A 444-75   Standard Specification for Steel Sheet, Zinc Coated (Galvanized) by the Hot Dip Process for Culverts and Underdrains

A 615-76a Standard Specification for Deformed and Plain Billet-Steel Bars for Concrete Reinforcement

A 616-76   Standard Specification for Rail-Steel Deformed and Plain Bars for Concrete Reinforcement

A 617-76   Standard Specification for Axle-Steel Deformed and Plain Bars for Concrete Reinforcement

A 706-76   Standard Specification for Low-Alloy Steel Deformed Bars for Concrete Reinforcement

C 39-72   Standard Specification for Compressive Strength of Cylindrical Concrete Specimens

E 329-72   Standard Recommended Practice for Inspection

and Testing Agencies for Concrete, Steel, and Bituminous Materials as Used in Construction

1.4.3. *ACI standards*

301-72   (revised 1975) Specifications for Structural Concrete for Buildings

318-77   Building Code Requirements for Reinforced Concrete

322-71   Building Code Requirements for Structural Plain Concrete

1.4.4. *AWS standards*

01.1   Structual Welding Code

012.1   Reinforcing Steel Welding Code

1.5. Project Conditions

1.5.1. *Examination of site*—Visit (prior to submitting bid) to determine existing surface conditions.

1.5.2. *Subsurface data*—A subsurface investigation has been made by ____ . Logs of borings and test data are available for Contractor's information and for his interpretation as to soil and water conditions that may be encountered at the site. Logs and test data are not represented as complete description of the site soil and water information but only display what was found in borings at the indicated locations. Contractor has the right to obtain additional information, if necessary in his judgment.

1.5.3. *Existing underground utilities*—Locate all existing underground utilities and construction in the field by a qualified surveyor so as to determine any conflicts with the work. Should conflicts be determined, do not proceed with the work until the Architect/Engineer specifies method(s) to eliminate the conflict.

Section 2—Materials and Construction

Part 2.1. General

2.1.1. Description

This section covers requirements for materials and construction for end bearing drilled piers, and includes the following:

2.1.1.1. Excavation and casing, dewatering, gas testing, and probing.

2.1.1.2. Reinforcing steel.

2.1.1.3. Concrete.

2.1.2. Submittals

2.1.2.1. *Geotechnical Engineer*—Will submit test reports to Architect-Engineer and to Contractor concerning allowable service load bearing pressures, elevations, dimensions, and alignment.

2.1.2.2. *Contractor*—Submit the following:
   a. Reinforcing steel shop drawings.
   b. Certified mill test reports for reinforcing steel.
   c. Evidence that proposed materials and mix designs conform to all requirements of "Specifications for Structural Concrete for Buildings (ACI 301-72) (Revised 1975)." except as modified by these specifications.
   d. Detailed procedures for casing removal, if any.
   e. Detailed procedures for tremie concrete, if any.
   f. Notification to Architect/Engineer to permit in-place inspection of reinforcing steel prior to placing concrete.
   g. Testing laboratory reports for concrete tests during construction.
   h. Reports of actual location, alignment, elevations, and dimensions of drilled piers.
   i. Reports of materials quantities, if specified.

2.1.3. Quality Assurance
   2.1.3.1. *Geotechnical Engineer*—Will provide inspection of all phases of drilled pier construction and request additional soil or rock testing if needed.
      2.1.3.1.A. (ADSC)—Prompt inspection and approval of drilled foundations must be made by Owner's Authorized Agent prior to placement of materials. Subsequent Inspection of the placement of materials should be prompt to permit removal, as an option, in the event of rejection. Inspect from the top if practical. Where practical, all work should avoid any unnecessary risk of putting a man in the hole.
   2.1.3.2. *Contractor*
      a. Provide the services of a qualified surveyor for performing all surveys and layouts and to determine vertical and horizontal alignments.
      b. Protect reinforcing steel from contamination.
   2.1.3.3. *Testing laboratory*—Will provide services conforming to the requirements of ASTM E 329, for sampling, testing, inspection, and reporting with respect to casing, reinforcing, and concrete.

2.1.4. Construction Tolerances

2.1.4.1. Bottom elevations of drilled piers as shown are estimated from soil boring data. Geotechnical Engineer will determine actual final bearing level during excavation.

2.1.4.2. *Maximum permissible variation of location*—1/24th of shaft diameter or 3 in., whichever is less.

2.1.4.2.A. (ADSC)—Variation not to exceed 3 inches.

2.1.4.3. *Concrete shafts out of plumb*—Not more than 1.5 percent of the length nor exceeding 12.5 percent of shaft diameter or 15 in., whichever is less.

2.1.4.3.A. (ADSC)—Where penetration of rock is required, 2 percent vertical tolerance is allowed.

2.1.4.3.B. (ADSC)—Install battered drilled shafts within 5 percent of the length from specified inclination.

2.1.4.4. If the tolerance of Articles 2.1.4.2 and 2.1.4.3 are exceeded, furnish and pay for corrective design and construction that may be required.

2.1.4.5. *Concrete cut-off elevation tolerance*—Plus 1 in. to minus 3 in.

2.1.4.6. (ADSC)—Where drilling mud is used and cut-off elevation is below grade, these tolerances may be too stringent.

2.1.5. Delivery, Handling, and Storage of Permanent Casing

2.1.5.2. Deliver casing to site in undamaged condition.

2.1.5.2. Handle and protect casing to maintain round within ±2 percent.

Part 2.2. Materials

2.2.1. Steel Casing

2.2.1.1. ASTM A 252, Grade 2, or ASTM A 36, or ASTM A 444 corrugated steel, as specified, or as shown on the contract drawings.

2.2.1.2. Furnish 100 percent penetration welds for vertical joints in noncorrugated permanent casings.

2.2.1.3. For permanent casing requiring hardened steel teeth for seating into rock, face weld teeth with AWS electrodes.

2.2.2. *Reinforcing steel*—ASTM A 615, A 616, A 617, or A 706, as specified, or as shown on the contract drawings.

2.2.3. *Concrete*—Concrete work shall conform to all requirements of "Specifications for Structural Concrete for Buildings (ACI 301-72) (Revised 1975)," except the following:

Sections 3.83, 3.84,   Chapter 9

and 3.85   Chapter 10

Chapter 4   Chapter 11

Section 5.4   Chapter 12

Sections 6.2, 6.3,   Chapter 13

6.4, and 6.5   Chapter 14

Section 7.4   Chapter 15

2.2.4. *Sand-cement grout*—As specified for project, for filling annular void outside permanent casing.

Part 2.3. Construction

2.3.1. Excavation, Soil Testing, and Inspection

2.3.1.1. Excavate drilled piers to dimensions and required elevations shown on contract drawings. Maintain sidewall stability during drilling. If drawings call for an allowable service load bearing pressure, extend excavation to suitable material.

2.3.1.2. Determine suitability of supporting material for drilled piers, as follows:

a. Explore bearing stratum to depth equal to the diameter of the bearing area below the bottom of the drilled pier with probe hole when directed by the Geotechnical Engineer.

b. Inspection and testing at the bottom of each pier will be by the Geotechnical Engineer.

c. Excavate for drilled pier bells (if required) immediately upon confirmation of the allowable service load bearing value by the Geotechnical Engineer.

d. If test results indicate the stratum is not capable of providing the required service load bearing pressure, notify the Architect-Engineer for a determination of adjustments to be made. These may include, but not be limited to, advancing the shaft length as directed by the Geotechnical Engineer and repeating the above steps, or enlarging the bell diameter as determined by the Architect-Engineer for the appropriate bearing pressure as determined by the Geotechnical Engineer.

2.3.1.3. Provide gas testing equipment, protective cage, or temporary casing of proper diameter, length, and thickness and other safety equipment called for by law for inspection and testing of drilled piers and to protect workmen during hand belling or other operations necessitating entry into shaft.

2.3.1.4. Check each drilled pier for toxic and explosive gases prior to personal entering. If gas is found, ventilate with forced air until safe for entry.

2.3.1.5. Remove from bottoms of drilled piers, loose material or free water in quantities sufficient to cause settlement or affect concrete strength as determined by the Geotechnical Engineer. Excavate pier bottoms to a level plane. If bottoms are sloping rock, excavate to a level plane or step with maximum step height less than one-quarter the width or diameter of the bearing area.

2.3.1.6. Remove excavated material from site or as otherwise directed by the Architect/Engineer.

2.3.2. Steel Casing

2.3.2.1. Provide steel casing for shaft excavation where required. Provide casing of sufficient strength to withstand handling stresses, concrete pressure, and surrounding earth and/or fluid pressures. Make diameter of excavation in relation to diameter of casing, such as to create a minimum of void space outside of casing. Provide permanent casing with minimum outside diameter equal to nominal outside diameter of shaft.

2.3.2.1.A. (ADSC)—Provide temporary casing with minimum outside diameter equal to normal outside diameter of drilled foundations.

2.3.2.2. Casing may be removed at option of Contractor unless otherwise specified. If casing is removed during or after concreting, follow special requirements specified in Article 2.3.2.

2.3.3. Reinforcing Steel

2.3.3.1. Place reinforcement for drilled piers in accordance with the contract documents.

2.3.3.2. Use reinforcement at time of placement which is free of mud, oil, or other coatings that adversely affect bond.

2.3.3.3. Reinforcement with rust, scale, or a combination of both may be used provided the minimum dimensions, including height of deformations and weight of wire-brushed specimens, are not less than required by applicable ASTM specifications. Architect/Engineer will determine acceptablity of such reinforcement.

2.3.3.4. Use metal reinforcement without kinks or non-specified bends. Straighten or repair bars in a manner that will not damage the bars or adjacent construction.

2.3.3.5. Place bars as shown on contract drawings with cover of not less than 3 in. where exposed to soil.

2.3.3.6. Make splices in reinforcement as shown on contract drawings unless otherwise accepted.

2.3.3.7. Provide clear distance between bars of not less than one and one-half times the bar diameter, nor one and one-half times the maximum aggregate size

2.3.3.7.A. (ADSC)—Reinforcing Steel (if required)

a. A reinforcing cage shall be designed as a structural element and braced to retain its configuration throughout the placing of concrete and the extraction of the casing from the shaft. Job limitations may require some reinforcing to be assembled in place.

b. The longitudinal rebar area required by the design shall be made up of bars of the largest practical size, to increase the rigidity of the cage. The horizontal steel shall be either spiral caging or a series of horizontal hoops, at the designer's option. Where spiral hooping or lateral ties are used, spacing shall not be less than 6″. Longitudinal rebars shall have a minimum spacing of 3″.

2.3.4. Concrete

    2.3.4.1. Dewater drilled pier excavation prior to placing concrete. Perform pumping in a manner that will not create ground loss problems that might adversely affect this and existing adjacent structures as determined by the Geotechnical Engineer. If during pumping excessive water inflow is noted, use alternative means to reduce inflow such as extending casing, outside deep wells, or grouting, or other acceptable means. If water seepage still is considered by the Geotechnical Engineer to be excessive for safe removal, follow procedure specified in Article 2.3.4.9.

    2.3.4.2. Obtain permission of Architect/Engineer prior to placing concrete.

        2.3.4.2.A. (ADSC)—If a water problem exists, the drilling contractor shall be permitted, at his option, to go directly to the underwater concrete placement method.

    2.3.4.3. Place concrete immediately after completion of excavation and after Geotechnical Engineer has verified allowable service load bearing capacity. Do not leave uncased or belled excavations open overnight.

    2.3.4.4. Free-fall concrete may be used provided it is directed through a hopper, or equivalent, such that fall is vertical down center of shaft without hitting sides or reinforcing. Vibrate top 5 ft of concrete, but only after casing has been pulled or when casing is permanent.

    2.3.4.5. Place concrete in pier in one continuous operation. If a construction joint is unavoidable, level, roughen, and clean surface prior to recommencement of concrete placement. Provide reinforcing dowels or a shear key when required by the Architect-Engineer.

    2.3.4.6. If casing is withdrawn, the Geotechnical Engineer will provide inspection during the removal of casing and placing of concrete. Withdraw casing only as shaft is filled with concrete. Maintain adequate head of concrete to balance outside soil and water pressure above the bottom of the casing at all times during withdrawal.

Specific procedures that the Contractor will follow to accomplish this objective shall be submitted for approval.

2.3.4.7. Where casing is removed, provide specially designed concrete with a minimum slump of 5 in. and with a retarder to prevent arching of concrete (during casing pulling) or setting of concrete until after casing is pulled. Check concrete level prior to, during, and after pulling casing. Avoid vibrating concrete if casing is pulled. Pull casing before slump decreases below 5 in. as determined by testing.

2.3.4.7.A. (ADSC)—During casing extraction, upward movement of the steel should not exceed 6 in. Downward movement should not exceed 6 in. per 20 feet shaft length.

2.3.4.8. When casing is left in place, fill void space between casing and shaft excavation with concrete or fluid grout by means of grout pipe and pump pressure as required.

2.3.4.9. For placing concrete under water, where permitted, use tremie pipe or concrete pumping with special procedures as specified or accepted.

2.3.4.10. *Concrete tests*—Take one set of four cylinders per drilled pier but not more than one set per truck, nor less than required by ACI 318-77. Test one sample at 7 days and two at 28 days; keep one sample in reserve for testing in the event of a low break.

Appendix D

Areas and Volumes of Shafts and Underreams

(This tabulation reprinted courtesy of Watson Incorporated, Fort Worth, TX)

BASE AREA IN SQUARE FEET
per Bell Diameter Inches

Bell Diameter Inches	Area	Bell Diameter Inches	Area
20	2.18	100	54.54
22	2.64	102	56.75
24	3.14	104	58.99
26	3.69	106	61.28
28	4.28	108	63.62
30	4.91	110	66.00
32	5.59	112	68.42
34	6.31	114	70.88
36	7.07	116	73.39
38	7.88	118	75.94
40	8.73	120	78.54
42	9.62	122	81.18
44	10.56	124	83.86
46	11.54	126	86.59
48	12.57	128	89.36
50	13.64	132	95.03
52	14.75	136	100.88
54	15.90	140	106.90
56	17.10	144	113.10
58	18.35	148	119.47
60	19.63	152	126.01
62	20.97	156	132.73
64	22.34	160	139.63
66	23.76	162	143.14
68	25.22	164	146.69
70	26.73	168	153.94
72	28.27	172	161.36
74	29.87	176	168.95
76	31.50	180	176.71
78	33.18	184	184.66
80	34.91	188	192.77
82	36.67	192	201.66
84	38.48	196	209.53
86	40.34	200	218.17
88	42.24	204	226.98
90	44.18	208	235.97
92	46.16	212	245.13
94	48.19	216	254.47
96	50.27		
98	52.38		

BASE AREA IN SQUARE METERS
per Bell Diameter Centimeters

Bell Diameter Centimeters	Area	Bell Diameter Centimeters	Area
50	.20	230	4.15
60	.28	240	4.52
70	.38	250	4.91
80	.50	260	5.31
90	.64	270	5.73
100	.79	280	6.16
110	.95	290	6.61
120	1.13	300	7.07
130	1.33	310	7.55
140	1.54	320	8.04
150	1.77	330	8.55
160	2.01	340	9.08
170	2.27	350	9.62
180	2.54	360	10.18
190	2.84	370	10.75
200	3.14	380	11.34
210	3.46	390	11.95
220	3.80	400	12.57
		410	13.20
		420	13.85
		430	14.52
		440	15.21
		450	15.90
		460	16.62
		470	17.35
		480	18.10
		490	18.86
		500	19.63
		510	20.43
		520	21.24
		530	22.06
		540	22.90
		550	23.76
		560	24.63
		570	25.52
		580	26.42
		590	27.34
		600	28.27

Bearing Area

Shaft Areas and Volumes

	Per Lineal Foot				Per Lineal Meter		
Shaft Diameter (in.)	Volume (cu yd)	Side Shear Area (sq ft)	Bearing Area (sq ft)	Shaft Diameter (cm)	Volume (m³)	Side Shear Area (m²)	Bearing Area (m²)
12	0.03	3.14	0.79	30	0.07	0.94	0.07
14	0.04	3.67	1.07	35	0.10	1.10	0.10
16	0.05	4.19	1.40	40	0.13	1.26	0.13
18	0.07	4.71	1.77	45	0.16	1.41	0.16
20	0.08	5.24	2.18	50	0.20	1.57	0.20
22	0.10	5.76	2.64	55	0.24	1.73	0.24
24	0.12	6.28	3.14	60	0.28	1.88	0.28
26	0.14	6.81	3.69	65	0.33	2.04	0.33
28	0.16	7.33	4.28	70	0.38	2.20	0.38
30	0.18	7.85	4.91	75	0.44	2.36	0.44
32	0.21	8.38	5.59	80	0.50	2.51	0.50
34	0.23	8.90	6.31	85	0.57	2.67	0.57
36	0.26	9.42	7.07	90	0.64	2.83	0.64
38	0.29	9.95	7.88	95	0.71	2.98	0.71
40	0.32	10.47	8.73	100	0.79	3.14	0.79
42	0.36	11.00	9.62	105	0.87	3.30	0.87
44	0.39	11.52	10.56	110	0.95	3.46	0.95
46	0.43	12.04	11.54	115	1.04	3.61	1.04
48	0.47	12.57	12.57	120	1.13	3.77	1.13

(Table continues on p. 216.)

Shaft Areas and Volumes (*Continued*)

Per Lineal Foot				Per Lineal Meter			
Shaft Diameter (in.)	Volume (cu yd)	Side Shear Area (sq ft)	Bearing Area (sq ft)	Shaft Diameter (cm)	Volume (m³)	Side Shear Area (m²)	Bearing Area (m²)
50	0.51	13.09	13.64	125	1.23	3.93	1.23
52	0.55	13.61	14.75	130	1.33	4.08	1.33
54	0.59	14.14	15.90	135	1.43	4.24	1.43
56	0.63	14.66	17.10	140	1.54	4.40	1.54
58	0.68	15.18	18.35	145	1.65	4.56	1.65
60	0.73	15.71	19.63	150	1.77	4.71	1.77
62	0.78	16.23	20.97	155	1.89	4.87	1.89
64	0.83	16.76	22.34	160	2.01	5.03	2.01
66	0.88	17.28	23.76	165	2.14	5.18	2.14
68	0.93	17.80	25.22	170	2.27	5.34	2.27
70	0.99	18.33	26.73	175	2.41	5.50	2.41
72	1.05	18.85	28.27	180	2.54	5.65	2.54
74	1.11	19.37	29.87	185	2.69	5.81	2.69
76	1.17	19.90	31.50	190	2.84	5.97	2.84
78	1.23	20.42	33.18	195	2.99	6.13	2.99
84	1.43	21.99	38.48	210	3.46	6.60	3.46
90	1.64	23.56	44.18	225	3.98	7.07	3.98
96	1.86	25.13	50.27	240	4.52	7.54	4.52
102	2.10	26.70	56.75	255	5.11	8.01	5.11
108	2.36	28.27	63.62	270	5.73	8.48	5.73
114	2.63	29.85	70.88	285	6.38	8.95	6.38
120	2.91	31.42	78.54	300	7.07	9.42	7.07
126	3.21	32.99	86.59	315	7.79	9.90	7.79
132	3.52	34.56	95.03	330	8.55	10.37	8.55

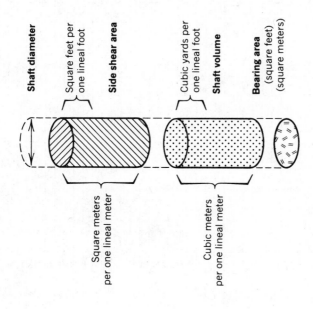

Shaft diameter

Square feet per one lineal foot

Side shear area

Cubic yards per one lineal foot

Shaft volume

Bearing area
(square feet)
(square meters)

Square meters per one lineal meter

Cubic meters per one lineal meter

BELL HEIGHT IN FEET

60°

Shaft Diameter Inches

Bell Diameter Inches	72	60	54	48	42	36	30	24	18
20									48
22									62
24									77
26								48	91
28								62	1.06
30								77	1.20
32							48	91	1.34
34							62	1.06	1.49
36							77	1.20	1.63
38						48	91	1.34	1.78
40						62	1.06	1.49	1.92
42						77	1.20	1.63	2.07
44					48	91	1.34	1.78	2.21
46					62	1.06	1.49	1.92	2.35
48					77	1.20	1.63	2.07	2.50
50				64	91	1.34	1.78	2.21	2.64
52				79	1.06	1.49	1.92	2.35	2.79
54				93	1.20	1.63	2.07	2.50	2.93
56			64	1.08	1.34	1.78	2.21	2.64	
58			79	1.22	1.49	1.92	2.35	2.79	
60			93	1.37	1.63	2.07	2.50	2.93	
62		64	1.08	1.51	1.78	2.21	2.64	3.08	
64		79	1.22	1.65	1.92	2.35	2.79	3.22	
66		93	1.37	1.80	2.07	2.50	2.93	3.36	
68		1.08	1.51	1.94	2.21	2.64	3.08	3.51	
70		1.22	1.65	2.09	2.35	2.79	3.22	3.65	
72		1.37	1.80	2.23	2.50	2.93	3.36	3.80	
74	64	1.51	1.94	2.38	2.64	3.08	3.51		
76	79	1.65	2.09	2.52	2.79	3.22	3.65		
78	93	1.80	2.23	2.67	2.93	3.36	3.80		
80	1.08	1.94	2.38	2.81	3.08	3.51	3.94		
82	1.22	2.09	2.52	2.95	3.22	3.65	4.09		
84	1.37	2.23	2.67	3.10	3.36	3.80	4.23		
86	1.51	2.38	2.81	3.24	3.51	3.94	4.37		
88	1.65	2.52	2.95	3.39	3.65	4.09	4.52		
90	1.80	2.67	3.10	3.53	3.80	4.23	4.66		
92	1.94	2.81	3.24	3.68	3.94	4.37			

45°

Shaft Diameter Inches

Bell Diameter Inches	72	60	54	48	42	36	30	24	18
20									42
22									50
24									58
26								42	67
28								50	75
30								58	83
32							42	67	92
34							50	75	1.00
36							58	83	1.08
38						42	67	92	1.17
40						50	75	1.00	1.25
42						58	83	1.08	1.33
44					42	67	92	1.17	1.42
46					50	75	1.00	1.25	1.50
48					58	83	1.08	1.33	1.58
50				58	67	92	1.17	1.42	1.67
52				67	75	1.00	1.25	1.50	1.75
54				75	83	1.08	1.33	1.58	1.83
56			58	83	92	1.17	1.42	1.67	
58			67	92	1.00	1.25	1.50	1.75	
60			75	1.00	1.08	1.33	1.58	1.83	
62		58	83	1.08	1.17	1.42	1.67	1.92	
64		67	92	1.17	1.25	1.50	1.75	2.00	
66		75	1.00	1.25	1.33	1.58	1.83	2.08	
68		83	1.08	1.33	1.42	1.67	1.92	2.17	
70		92	1.17	1.42	1.50	1.75	2.00	2.25	
72	58	1.00	1.25	1.50	1.58	1.83	2.08	2.33	
74	67	1.08	1.33	1.58	1.67	1.92	2.17		
76	75	1.17	1.42	1.67	1.75	2.00	2.25		
78	83	1.25	1.50	1.75	1.83	2.08	2.33		
80	92	1.33	1.58	1.83	1.92	2.17	2.42		
82	92	1.42	1.67	1.92	2.00	2.25	2.50		
84	1.00	1.50	1.75	2.00	2.08	2.33	2.58		
86	1.08	1.58	1.83	2.08	2.17	2.42	2.67		
88	1.17	1.67	1.92	2.17	2.25	2.50	2.75		
90	1.25	1.75	2.00	2.25	2.33	2.58	2.83		
92	1.33	1.83	2.08	2.33	2.42	2.67			

Shaft diameter inches

Bell diameter inches — Table 1

Bell dia	18	24	30	36	42	48	54	60	72
94				2.75	2.50	2.42	2.17	1.92	1.42
96				2.83	2.58	2.50	2.25	2.00	1.50
98				2.92	2.67	2.58	2.33	2.08	1.58
100				3.00	2.75	2.67	2.42	2.17	1.67
102				3.08	2.83	2.75	2.50	2.25	1.75
104				3.17	2.92	2.83	2.58	2.33	1.83
106				3.25	3.00	2.92	2.67	2.42	1.92
108				3.33	3.08	3.00	2.75	2.50	2.00
110					3.17	3.08	2.83	2.58	2.08
112					3.25	3.17	2.92	2.67	2.17
114					3.33	3.25	3.00	2.75	2.25
116					3.42	3.33	3.08	2.83	2.33
118					3.50	3.42	3.17	2.92	2.42
120					3.58	3.50	3.25	3.00	2.50
122					3.67	3.58	3.33	3.08	2.58
124					3.75	3.67	3.42	3.17	2.67
126					3.83	3.75	3.50	3.25	2.75
128						3.83	3.58	3.33	2.83
132						4.00	3.75	3.50	3.00
136						4.17	3.92	3.67	3.17
140						4.33	4.08	3.83	3.33
144						4.50	4.25	4.00	3.50
148							4.42	4.17	3.67
152							4.58	4.33	3.83
156							4.75	4.50	4.00
160							4.92	4.67	4.17
162							5.00	4.75	4.25
164								4.83	4.33
168								5.00	4.50
172								5.17	4.67
176								5.33	4.83
180								5.50	5.00
184									5.17
188									5.33
192									5.50
196									5.67
200									5.83
204									6.00
208									6.17
212									6.33
216									6.50

Bell diameter inches — Table 2

Bell dia	18	24	30	36	42	48	54	60	72
94				4.52	4.09	3.82	3.39	2.95	2.09
96				4.66	4.23	3.96	3.53	3.10	2.23
98				4.81	4.37	4.11	3.68	3.24	2.38
100				4.95	4.52	4.25	3.82	3.39	2.52
102				5.10	4.66	4.40	3.96	3.53	2.67
104				5.24	4.81	4.54	4.11	3.68	2.81
106				5.39	4.95	4.69	4.25	3.82	2.95
108				5.53	5.10	4.83	4.40	3.96	3.10
110					5.24	4.97	4.54	4.11	3.24
112					5.39	5.12	4.69	4.25	3.39
114					5.53	5.26	4.83	4.40	3.53
116					5.67	5.41	4.97	4.54	3.68
118					5.82	5.55	5.12	4.69	3.82
120					5.96	5.70	5.26	4.83	3.96
122					6.11	5.84	5.41	4.97	4.11
124					6.25	5.98	5.55	5.12	4.25
126					6.40	6.13	5.70	5.26	4.40
128						6.27	5.84	5.41	4.54
132						6.56	6.13	5.70	4.83
136						6.85	6.42	5.98	5.12
140						7.14	6.71	6.27	5.41
144						7.43	7.00	6.56	5.70
148							7.28	6.85	5.98
152							7.57	7.14	6.27
156							7.86	7.43	6.56
160							8.15	7.72	6.85
162							8.29	7.86	7.00
164								8.01	7.14
168								8.29	7.43
172								8.58	7.72
176								8.87	8.01
180								9.16	8.29
184									8.58
188									8.87
192									9.16
196									9.45
200									9.74
204									10.03
208									10.31
212									10.60
216									10.89

Diagram labels: Bell Height; Toe Height; 60° or 45°.

For Shaft Diameters less than 42 inches, a 4 inch bell ''toe'' is used in calculations.   Bell Diameter is calculated to 3 times the size of the Shaft Diameter.

For Shaft Diameters greater than 42 inches, a 6 inch bell ''toe'' is used in calculations.

219

BELL VOLUME IN CUBIC YARDS

60°

Shaft Diameter Inches

Bell Diameter Inches	18	24	30	36	42	48	54	60	72
20	.006								
22	.015								
24	.027	.008							
26	.043	.020							
28	.062	.035							
30	.085	.055	.010						
32	.112	.079	.025						
34	.143	.107	.043						
36	.179	.140	.067	.012					
38	.219	.178	.096	.029					
40	.265	.222	.130	.051					
42	.315	.270	.169	.079	.014				
44	.371	.325	.214	.113	.034				
46	.433	.385	.265	.152	.059				
48	.501	.451	.322	.197	.091	.023			
50	.574	.524	.385	.249	.129	.052			
52	.655	.604	.455	.307	.174	.088			
54	.741	.690	.532	.373	.226	.131	.025		
56		.783	.615	.445	.284	.182	.058		
58		.884	.707	.525	.350	.240	.098		
60		.992	.805	.612	.424	.306	.147	.028	
62			.912	.707	.505	.380	.203	.064	
64		1.108	1.026	.810	.594	.462	.267	.109	
66		1.232	1.149	.921	.692	.553	.340	.162	
68		1.364	1.280	1.040	.798	.653	.422	.224	
70		1.505	1.420	1.169	.913	.761	.512	.294	
72		1.654	1.569	1.306	1.036	.879	.612	.374	.034
74			1.727	1.452	1.169	1.006	.722	.464	.077
76			1.894	1.608	1.311	1.143	.841	.563	.130
78			2.071	1.774	1.463	1.290	.969	.672	.192
80			2.258	1.949	1.625	1.446	1.108	.791	.265
82			2.455	2.134	1.796	1.614	1.258	.920	.349
84			2.662	2.330	1.979	1.791	1.418	1.060	.442
86			2.880	2.537	2.171	1.980	1.589	1.211	.547
88			3.109	2.754	2.375	2.180	1.771	1.373	.663
90				2.982	2.589	2.391	1.964	1.546	.790

45°

Shaft Diameter Inches

Bell Diameter Inches	18	24	30	36	42	48	54	60	72
20	.006								
22	.013								
24	.023								
26	.035	.008							
28	.049	.017							
30	.065	.030	.009						
32	.085	.045	.021						
34	.106	.062	.036						
36	.131	.082	.054	.011					
38	.158	.106	.075	.025					
40	.189	.132	.099	.043					
42	.223	.162	.127	.064	.013				
44	.260	.195	.158	.088	.030				
46	.301	.231	.193	.116	.050				
48	.345	.271	.231	.148	.074	.021			
50	.394	.315	.274	.184	.101	.047			
52	.446	.363	.320	.224	.133	.077			
54	.502	.415	.371	.268	.169	.111	.024		
56		.471	.425	.316	.210	.150	.053		
58		.532	.485	.368	.255	.194	.086		
60		.596	.549	.426	.304	.242	.124		
62		.666	.617	.488	.358	.296	.167	.027	
64		.740	.691	.555	.417	.355	.216	.058	
66		.819	.769	.626	.481	.418	.269	.095	
68		.903	.853	.703	.550	.488	.328	.137	
70		.992	.942	.785	.624	.550	.393	.185	
72		1.086	1.036	.873	.704	.562	.463	.238	
74			1.136	.966	.789	.643	.539	.296	.032
76			1.241	1.065	.880	.729	.621	.361	.070
78			1.352	1.169	.977	.821	.708	.431	.113
80			1.469	1.280	1.079	.919	.803	.507	.163
82			1.593	1.396	1.188	1.024	.903	.590	.219
84			1.722	1.519	1.303	1.134	1.010	.679	.281
86			1.858	1.648	1.424	1.252	1.123	.774	.350
88			2.000	1.784	1.551	1.375	1.244	.876	.425
90				1.926	1.686	1.506	1.371	.985	.507
92						1.643		1.100	.596

60° **45°**

Shaft Diameter Inches

Bell Diameter Inches

Bell Volume 60 or 45

Bell Dia.	60° 18	24	30	36	42	48	54	60	72	45° 18	24	30	36	42	48	54	60	72
94				3.221	2.815	2.613	2.169	1.731	929				2.074	1.827	1.788	1.505	1.223	692
96				3.472	3.052	2.848	2.385	1.928	1.079				2.230	1.974	1.939	1.646	1.353	795
98				3.735	3.301	3.094	2.614	2.136	1.241				2.393	2.129	2.098	1.794	1.490	905
100				4.010	3.562	3.353	2.854	2.357	1.416				2.563	2.291	2.264	1.950	1.634	1.023
102				4.297	3.836	3.624	3.108	2.590	1.603				2.740	2.460	2.438	2.114	1.786	1.148
104				4.596	4.121	3.908	3.373	2.836	1.803				2.924	2.637	2.619	2.285	1.946	1.281
106				4.908	4.420	4.204	3.652	3.095	2.016				3.116	2.821	2.809	2.464	2.113	1.422
108				5.233	4.731	4.514	3.944	3.367	2.242				3.316	3.013	3.006	2.651	2.288	1.571
110					5.055	4.837	4.250	3.653	2.481					3.213	3.211	2.846	2.472	1.728
112					5.393	5.174	4.569	3.952	2.734					3.420	3.425	3.049	2.663	1.893
114					5.744	5.525	4.902	4.265	3.001					3.636	3.646	3.260	2.863	2.066
116					6.109	5.890	5.249	4.592	3.282					3.860	3.877	3.480	3.072	2.248
118					6.489	6.269	5.610	4.933	3.577					4.092	4.116	3.709	3.289	2.438
120					6.882	6.663	5.986	5.289	3.887					4.333	4.363	3.946	3.515	2.637
122					7.290	7.072	6.377	5.660	4.211					4.586	4.620	4.193	3.750	2.845
124					7.713	7.495	6.783	6.046	4.551					4.841	4.885	4.448	3.993	3.063
126					8.151	7.934	7.204	6.447	4.906					5.108	5.160	4.712	4.246	3.289
128						8.389	7.640	6.864	5.276						5.444	4.986	4.509	3.524
132						9.345	8.561	7.745	6.065						6.041	5.562	5.061	4.024
136						10.366	9.546	8.609	6.918						6.676	6.177	5.653	4.562
140						11.454	10.598	9.702	7.838						7.351	6.831	6.285	5.140
144						12.610	11.719	10.783	8.826						8.067	7.527	6.957	5.760
148							12.910	11.934	9.885							8.264	7.672	6.421
152							14.173	13.158	11.016							9.045	8.429	7.125
156							15.511	14.455	12.221							9.870	9.231	7.873
160							16.924	15.829	13.502							10.740	10.078	8.667
162							17.160	16.545	14.172							11.192	10.519	9.081
164								17.280	14.861								10.971	9.507
168								18.811	16.300								11.912	10.394
172								20.424	17.821								12.901	11.330
176								22.120	19.424								13.940	12.316
180								23.902	21.114								15.029	13.352
184									22.890									14.440
188									24.756									15.580
192									26.712									16.775
196									28.761									18.024
200									30.905									19.329
204									33.146									20.692
208									35.484									22.113
212									37.924									23.592
216									40.465									25.133

BELL HEIGHT IN METERS

Shaft Diameter Centimeters

Bell Diameter Centimeters

60°

	40	60	80	100	120	140	160	180	200
50	.19								
60	.27								
70	.36	.19							
80	.45	.27							
90	.53	.36	.19						
100	.62	.45	.27						
110	.71	.53	.36	.19					
120	.79	.62	.45	.27					
130		.71	.53	.36	.24				
140		.79	.62	.45	.33				
150		.88	.71	.53	.41	.24			
160		.97	.79	.62	.50	.33			
170		1.05	.88	.71	.59	.41	.24		
180		1.14	.97	.79	.67	50	.33		
190			1.05	.88	.76	.59	.41	.24	
200			1.14	.97	.85	.67	.50	.33	
210			1.23	1.05	.93	.76	59	.41	.24
220			1.31	1.14	1.02	.85	.67	.50	.33
230			1.40	1.23	1.11	.93	.76	.59	.41
240			1.49	1.31	1.19	1.02	.85	.67	.50
250				1.40	1.28	1.11	.93	.76	.59
260				1.49	1.36	1.19	1.02	.85	.67
270				1.57	1.45	1.28	1.11	.93	.76
280				1.66	1.54	1.36	1.19	1.02	.85
290				1.75	1.62	1.45	1.28	1.11	.93

45°

	40	60	80	100	120	140	160	180	200
50	.15								
60	.20								
70	.25	.15							
80	.30	.20							
90	.35	.25	.15						
100	.40	.30	.20						
110	.45	.35	.25	.15					
120	.50	.40	.30	.20					
130		.45	.35	.25	20				
140		.50	.40	.30	.25				
150		.55	.45	.35	.30	.20			
160		.60	.50	.40	.35	.25			
170		65	.55	.45	.40	.30	.20		
180		.70	.60	.50	.45	.35	.25		
190			.65	.55	.50	.40	.30	.20	
200			.70	.60	.55	.45	.35	.25	
210			.75	.65	.60	.50	.40	.30	.20
220			.80	.70	.65	.55	.45	.35	.25
230			.85	.75	.70	.60	.50	.40	.30
240			.90	.80	.75	.65	.55	.45	.35
250				.85	.80	.70	.60	.50	40
260				.90	.85	.75	65	.55	45
270				.95	.90	.80	.70	.60	.50
280				1.00	.95	.85	.75	.65	.55
290				1.05	1.00	.90	.80	.70	60

Shaft Diameter Centimeters

Upper table

Bell Diameter	40	60	80	100	120	140	160	180	200
300				1.10	1.05	.95	.85	.75	.65
310					1.10	1.00	.90	.80	.70
320					1.15	1.05	.95	.85	.75
330					1.20	1.10	1.00	.90	.80
340					1.25	1.15	1.05	.95	.85
350					1.30	1.20	1.10	1.00	.90
360					1.35	1.25	1.15	1.05	.95
370						1.30	1.20	1.10	1.00
380						1.35	1.25	1.15	1.05
390						1.40	1.30	1.20	1.10
400						1.45	1.35	1.25	1.15
410						1.50	1.40	1.30	1.20
420						1.55	1.45	1.35	1.25
430							1.50	1.40	1.30
440							1.55	1.45	1.35
450							1.60	1.50	1.40
460							1.65	1.55	1.45
470							1.70	1.60	1.50
480							1.75	1.65	1.55
490								1.70	1.60
500								1.75	1.65
510								1.80	1.70
520								1.85	1.75
530								1.90	1.80
540								1.95	1.85
550									1.90
560									1.95
570									2.00
580									2.05
590									2.10
600									2.15

Lower table

Bell Diameter	40	60	80	100	120	140	160	180	200
300				1.83	1.71	1.54	1.36	1.19	1.02
310					1.80	1.62	1.45	1.28	1.11
320					1.88	1.71	1.54	1.36	1.19
330					1.97	1.80	1.62	1.45	1.28
340					2.06	1.88	1.71	1.54	1.36
350					2.14	1.97	1.80	1.62	1.45
360					2.23	2.06	1.88	1.71	1.54
370						2.14	1.97	1.80	1.62
380						2.23	2.06	1.88	1.71
390						2.32	2.14	1.97	1.80
400						2.40	2.23	2.06	1.88
410						2.49	2.32	2.14	1.97
420						2.58	2.40	2.23	2.06
430							2.49	2.32	2.14
440							2.58	2.40	2.23
450							2.66	2.49	2.32
460							2.75	2.58	2.40
470							2.84	2.66	2.49
480							2.92	2.75	2.58
490								2.84	2.66
500								2.92	2.75
510								3.01	2.84
520								3.10	2.92
530								3.18	3.01
540								3.27	3.10
550									3.18
560									3.27
570									3.36
580									3.44
590									3.53
600									3.62

Diagram labels: Bell Height, Toe Height, 60 or 45.

Bell Diameter Centimeters

BELL VOLUME IN CUBIC METERS

60°

Shaft Diameter Centimeters

	40	60	80	100	120	140	160	180	200
50	.010								
60	.029	.015							
70	.057	.040							
80	.096	.079	.019						
90	.148	.131	.052						
100	.214	.198	.101	.024					
110	.295	.282	.155	.064					
120	.392	.384	.248	.122					
130		.505	.350	.200	.038				
140		.647	.473	.298	.097				
150		.810	.618	.418	.177	.044			
160		.997	.786	.562	.279	.112			
170		1.209	.978	.730	.406	.203	.051		
180			1.197	.924	.558	.320	.127	.057	
190			1.443	1.146	.737	.464	.230	.142	
200			1.719	1.397	.945	.636	.361	.256	
210			2.024	1.678	1.182	.838	.522	.402	.063
220			2.361	1.990	1.450	1.070	.714	.580	.157
230			2.730	2.335	1.750	1.335	.938	.791	.283
240				2.715	2.084	1.634	1.196	1.038	.443
250				3.130	2.453	1.967	1.488	1.321	.638
260				3.582	2.859	2.338	1.818	1.642	.869
270				4.072	3.302	2.745	2.185	2.002	1.138
280				4.602	3.784	3.192	2.591	2.402	1.446
290				5.173	4.307	3.680	3.038	2.844	1.795
300					4.872	4.209	3.526	3.330	2.186
310					5.480	4.782	4.058	3.860	2.619
320					6.132	5.399	4.634	4.634	3.098

45°

Shaft Diameter Centimeters

	40	60	80	100	120	140	160	180	200
50	.009								
60	.023	.013							
70	.044	.033							
80	.072	.061	.017						
90	.108	.097	.042						
100	.152	.143	.077	.021					
110	.206	.199	.122	.052					
120	.270	.266	.179	.094					
130		.345	.247	.148	.035				
140		.437	.327	.214	.082				
150		.542	.421	.294	.143	.040			
160		.661	.529	.388	.218	.095	.046		
170		.795	.653	.497	.308	.164	.108		
180			.791	.622	.413	.250	.186	.051	
190			.947	.763	.536	.351	.281	.120	
200			1.119	.921	.675	.470	.395	.207	.057
210			1.310	1.098	.833	.680	.527	.313	.133
220			1.520	1.294	1.009	.764	.680	.438	.229
230			1.749	1.509	1.205	.939	.852	.584	.345
240				1.744	1.422	1.136	1.046	.752	.482
250				2.001	1.660	1.353	1.262	.941	.641
260				2.280	1.920	1.592	1.501	1.153	.824
270				2.582	2.202	1.855	1.763	1.388	1.029
280				2.907	2.509	2.141	2.050	1.648	1.259
290				3.256	2.839	2.451	2.362	1.934	1.515
300					3.195	2.786	2.699	2.245	1.796
310					3.577	3.148	3.064	2.582	2.104
320					3.985	3.536			

BELL VOLUME IN CUBIC METERS

 60°   **45°**

Shaft Diameter Centimeters

Bell Dia	60° 40	60	80	100	120	140	160	180	200	45° 40	60	80	100	120	140	160	180	200
50	.010									.009								
60	.029									.023								
70	.057	.015								.044	.013							
80	.096	.040								.072	.033							
90	.148	.079	.019							.108	.061	.017						
100	.214	.131	.052							.152	.097	.042						
110	.295	.198	.101	.024						.206	.143	.077	.021					
120	.392	.282	.165	.064						.270	.199	.122	.052					
130		.384	.248	.122	.038						.266	.179	.094	.035				
140		.505	.350	.200	.097						.345	.247	.148	.082				
150		.647	.473	.298	.177	.044					.437	.327	.214	.143	.040			
160		.810	.618	.418	.279	.112					.542	.421	.294	.218	.095			
170		.997	.786	.562	.406	.203	.051				.661	.529	.388	.308	.164	.046		
180		1.209	.978	.730	.558	.320	.127				.795	.653	.497	.413	.250	.108		
190			1.197	.924	.737	.464	.230	.057				.791	.622	.536	.351	.186	.051	
200			1.443	1.146	.945	.636	.361	.142				.947	.763	.675	.470	.281	.120	
210			1.719	1.397	1.182	.838	.522	.256	.063			1.119	.921	.833	.680	.395	.207	.057
220			2.024	1.678	1.450	1.070	.714	.402	.157			1.310	1.098	1.009	.764	.527	.313	.133
230			2.361	1.990	1.750	1.335	.938	.580	.283			1.520	1.294	1.205	.939	.680	.438	.229
240			2.730	2.335	2.084	1.634	1.196	.791	.443			1.749	1.509	1.422	1.136	.852	.584	.345
250				2.715	2.453	1.967	1.488	1.038	.638				1.744	1.660	1.353	1.046	.752	.482
260				3.130	2.859	2.338	1.818	1.321	.869				2.001	1.920	1.592	1.262	.941	.641
270				3.582	3.302	2.745	2.185	1.642	1.138				2.280	2.202	1.855	1.501	1.153	.824
280				4.072	3.784	3.192	2.591	2.002	1.446				2.582	2.509	2.141	1.763	1.388	1.029
290				4.602	4.307	3.680	3.038	2.402	1.795				2.907	2.839	2.451	2.050	1.648	1.259
300				5.173	4.872	4.209	3.526	2.844	2.186				3.256	3.195	2.786	2.362	1.934	1.515
310					5.480	4.782	4.058	3.330	2.619					3.577	3.148	2.699	2.245	1.796
320					6.132	5.399	4.634	3.860	3.098					3.985	3.536	3.064	2.582	2.104
330					6.830	6.062	5.256	4.436	3.622					4.422	3.951	3.456	2.948	2.440
340					7.576	6.772	5.926	5.059	4.193					4.886	4.395	3.876	3.342	2.803
350					8.370	7.531	6.644	5.731	4.813					5.379	4.868	4.326	3.764	3.196
360					9.214	8.340	7.412	6.452	5.484					5.903	5.371	4.805	4.217	3.619
370						9.200	8.231	7.226	6.205						5.905	5.315	4.700	4.073
380						10.113	9.103	8.052	6.979						6.470	5.857	5.215	4.558
390						11.080	10.030	8.932	7.808						7.067	6.431	5.762	5.075
400						12.103	11.011	9.867	8.692						7.698	7.037	6.342	5.625
410						13.182	12.050	10.859	9.632						8.362	7.678	6.956	6.209
420						14.319	13.147	11.910	10.631						9.061	8.353	7.605	6.828
430							14.303	13.020	11.689							9.064	8.289	7.482
440							15.520	14.191	12.808							9.810	9.009	8.172
450							16.799	15.424	13.990							10.594	9.766	8.899
460							18.142	16.721	15.235							11.415	10.560	9.664
470							19.550	18.082	16.545							12.275	11.393	10.467
480							21.025	19.510	17.921							13.175	12.266	11.310
490								21.006	19.365								13.179	12.193
500								22.571	20.878								14.132	13.116
510								24.206	22.462								15.127	14.082
520								25.913	24.117								16.165	15.090
530								27.693	25.845								17.246	16.141
540								29.548	27.648								18.371	17.236
550									29.527									18.375
560									31.483									19.561
570									33.518									20.793
580									35.632									22.072
590									37.828									23.399
600									40.106									24.774

Bell Diameter Centimeters

Bell Volume — 60 or 45

For Shaft Diameters **less** than 1 meter, a 10 cm bell "toe" is used in calculations.
For Shaft Diameters **greater** than 1 meter, a 15 cm bell "toe" is used in calculations.

Bell diameter is calculated to 3 times the size of shaft diameter.

Appendix E

Conversion Factors, English to Metric (International System of Units)

The following are conversion units for quantities in common use in the aspects of geotechnical engineering discussed in this book—plus a few other related units. For the reader interested in a clear and complete discussion and tabulation of the conversions to and use of the entire International System of Units (SI), the reader is referred to "Metric Practice Guide," ASTM Designation E 380-84, which can be obtained from the American Society for Testing and Materials, 1916 Race Street, Philadelphia, PA 19103.

It will be noted that the SI units are, for the most part, the "metric system" or "cgs" units that most readers are familiar with. The principal change is the introduction of the unit of force, the newton (N), taking the place of the kilogram-force (kgf); and the name kilogram by itself designates mass. Note also that the newton is a much smaller unit than the kgf; it takes 9.806650 newtons to equal 1 kgf.

Note also that the table that follows presents conversion factors to only three significant figures. This is sufficient for most foundation engineering practice. If greater precision is needed, the reader is referred to the "Metric Practice Guide" mentioned above.

To Convert from:	To:	Multiply by:
	AREA	
foot^2 (sq ft)	meter^2 (m^2)	9.29×10^{-2}
inch^2 (sq in.)	meter^2 (m^2)	6.45×10^{-4}
inch^2 (sq in.)	centimeter^2 (cm^2)	6.45

To Convert from:	To:	Multiply by:
	AREA	
yard^2 (sq yd)	meter^2 (m^2)	8.36×10^{-1}
	TORQUE OR BENDING MOMENT	
pound (force)-foot (lbf-ft)*	newton-meter (N-m)	1.36
pound (force)-foot (lbf-ft)	kilogram-force-meter (kgf-m)	1.38×10^{-1}
	FORCE	
kilogram-force (kgf)	newton (N)	9.81
kip (kilopound-force)	newton (N)	4.45×10^3
kip (kilopound-force)	kilogram-force (kgf)	4.54×10^2
pound-force (lbf)	newton	4.45
pound-force (lbf)	kilogram-force (kgf)	4.54×10^{-1}
	LENGTH	
foot (ft)	meter (m)	3.05×10^{-1}
foot (ft)	centimeter (cm)	3.05×10
inch (in.)	meter (m)	2.54×10^{-2}
inch (in.)	centimeter (cm)	2.54
yard (yd)	meter (m)	9.14×10^{-1}
	MASS	
pound-mass (lbm)	kilogram (kg)	4.54×10^{-1}
ton-mass (short, 2000 lbm)	kilogram (kg)	9.07×10^2
kip-mass	kilogram (kg)	4.54×10^2
	MASS/VOLUME (DENSITY)	
pound-mass/foot^3†	kilogram/meter^3 (kg/m^3)	1.60×10
pound-mass/gallon (U.S.)	kilogram/meter^3 (kg/m^3)	1.20×10^2
pound-mass/cubic yard	kilogram/meter^3 (kg/m^3)	5.93×10^{-1}

* Most auger machine manufacturers list their torque ratings as "foot-pounds." This is incorrect. Foot-pounds are units of energy or work, not torque.
† Note that measurements of density of any material are measurements of mass per unit volume—not weight (force). A pound and a ton are units of force; a pound-mass is the mass that would weigh one pound on earth at sea level. A kilogram is a unit of mass, convertible to pounds mass.

To Convert from:	To:	Multiply by:
PRESSURE OR STRESS (FORCE/AREA)		
atmosphere (normal)	newton/meter^2 (N/m^2)	1.01×10^3
foot of water	newton/meter^2 (N/m^2)	2.99×10^3
kilogram-force/ centimeter^2 (kgf/cm^2)	newton/meter^2 (N/m^2)	9.81×10^4
kip/foot^2 (ksf)	newton/meter^2 (N/m^2)	4.79×10^4
pound-force/foot^2 (psf)	newton/meter^2	4.79×10
pound-force/inch^2 (psi)	newton/meter^2 (N/m^2)	6.89×10^3
pound-force/inch^2 (psi)	kilogram-force/ centimeter^2 (kgf/cm^2)	7.03×10^{-2}
ton-force/ft^2 (tsf)	newton/meter^2 (N/m^2)	9.58×10^4
ton/foot^2* (tsf)	kilogram-force/ centimeter^2 (kgf/ cm^2)*	9.77×10^{-1}

* These quantities are usually taken as equivalent in geotechnical engineering practice because field and laboratory measurements are rarely precise enough to warrant making a distinction.

	VOLUME	
foot^3	meter^3 (m^3)	2.83×10^{-2}
gallon (U.S. liquid)	meter^3 (m^3)	3.79×10^{-3}
gallon (U.S. liquid)	liter	3.79
inch^3 (cu in.)	meter^3 (m^3)	1.64×10^{-5}
liter	meter^3 (m^3)	1.00×10^{-3}
liter	gallon (U.S. liquid)	2.64×10^{-1}

Appendix F

Glossary of Drilling and Foundation Terms

Adhesion The property of a substance (in our case, cohesive soil) to "stick," "cling," or "adhere" to a solid structural element such as a concrete pier or pile, and thus establish a resistance to shearing movement between the soil mass and the structural element.

Adobe A term applied to a great variety of light-colored soils, ranging from clayey sandy silts to very plastic clays, in the southwestern United States.

ADSC Association of Drilled Shaft Contractors (an international association of foundation drilling contractors), address P.O. Box 75228, Dallas, TX 75228.

Aggregate The stone used in making concrete. "Fine aggregate" is sand; "coarse aggregate," gravel or gravel-size crushed stone.

Allowable Load The load which cannot be exceeded without incurring (in the opinion of the designer) risk of damaging structural movement.

Anchor Pier A pier designed to resist uplift or lateral forces.

Artesian Water Subsurface water which has sufficient pressure to raise the water in wells above the existing ground surface.

ASFE Association of Soil and Foundation Engineers.

ASTM American Society for Testing and Materials.

Auger A helical rotary tool for drilling a cylindrical hole in soil and/or rock.

Axial Load That portion of the load on a pier or pile which is in the direction of its axis.

Backfill   Any material placed in an excavated area, for the purpose of raising the grade in the area.

Bad Air   Contaminated air or gas, which constitutes a hazard to persons inspecting or working in an excavation. Sources of bad air include marsh gas, garbage fills, hazardous waste dumps, leaking gas or petroleum products, volcanic sources, naturally occurring petrochemical substances, or carbon monoxide. "Bad air" may also include air with insufficient oxygen to sustain respiration of personnel.

Bailing Bucket   A bucket-like tool for removing water from the hole during drilling or in preparation for concrete placement.

Batter   Angle with the vertical, normally expressed as a ratio of horizontal to vertical (i.e., 1:4 = 1 horizontal to 4 vertical).

Bearing Stratum   A soil or rock stratum which is expected to carry the pier load (either by end bearing or by sidewall friction, or by a combination of the two).

Bell   Enlargement of the lower end of a pier excavation, to increase the bearing area of the pier. (Also called "underream".)

Belling Bucket–Underreaming Bucket   A drilling bucket tool with expanding cutters that can enlarge the bottom of the drilled hole, to form a bell or underream. See **Bucket Auger, Drilling Bucket**.

Bentonite   The mineral, sodium montmorillonite, a highly expansive colloidal clay; the basis of "driller's mud." Suspended in a water slurry, it is used to prevent collapse of holes augered into cohesionless soils. Commercially sold as "Aquagel."

Boulder   A rock, usually rounded by weathering and abrasion, greater than 8 in. in size.

Bucket Auger (or Drilling Bucket)   A cylindrical rotary drilling tool with a hinged bottom containing a soil cutting blade; spoil enters the "bucket" and is lifted out of the hole, swung aside, and dumped by releasing the latch on the hinged bottom.

"Bull's Liver"   A local term in the northeastern United States for inorganic silt below the water table (a very unstable, "quaky" material).

Cage   Reinforcing bars preassembled for quick placing in a pier or excavated slurry-wall trench. (Rebar cage).

Caisson   Structural chamber used to keep soil and water from entering into a deep excavation or construction area. Caissons may be prefabricated and installed by being sunk in place by adding weight or by systematically excavating below the bottom of the unit to allow it to sink to the desired depth. In some areas of North America the term "caisson" is used to designate a cased or uncased drilled shaft.

Cake (Filter Cake)   A layer of clay or clayey soil, built up on the wall of a boring drilled with slurry (drilling mud, bentonite, etc.), having the effect of forming an impermeable lining to prevent (or diminish) loss of water from the hole, and maintain slurry pressure against the wall of the hole.

Caliche A soil cemented by carbonates left by evaporating groundwater; common in southwestern United States.

Calyx (or Shot) Barrel A core barrel without hard-metal cutting teeth, with which the rock is cut (or ground up) by chilled steel shot which roll and are ground up under the rotating steel edge of the barrel.

Capillarity The upward movement of water, due to effects of wetting and surface tension, that occurs through the very small void spaces that exist in a soil mass.

Casing An open-end steel pipe installed by drilling, driving or vibrating; to support the wall of a hole; to seal out groundwater; or to protect the concrete of the shaft from contamination by sloughing of the sides of the hole.

Caving (or Sloughing) Soil A soil which tends to fall into an uncased hole, during or after the drilling. Usually a cohesionless soil.

Changed Conditions Job conditions which differ substantially from conditions as represented in the plans and specifications, and/or the contract documents.

Chemical Grout A low-viscosity grout comprised of chemical compounds in liquid form or in a liquid solution (not solids suspended in water), which can be injected into soil or fissured rock to serve as a water cutoff or to improve load-carrying capacity.

Chicago Caisson A 3-ft or larger-diameter shaft with vertical timber sheathing, sunk in increments as circular bracing and additional sheathing are installed. May also apply to rectangular shafts.

Clay Cohesive soils which are firmly coherent, compact and hard when dry, but usually stiff, viscid, and ductile when moist. Smooth to the touch. Clays shrink on drying and expand on taking on moisture.

CMP (Corrugated Metal Pipe) See **Liner**.

Coarse-Grained Soil The soil types which have particles large enough to be seen without magnification. The coarse-grained soils include the sand and gravel (or larger) soil particles.

Cobble A rock fragment, 3 to 8 inches in size, usually rounded by weathering and abrasion.

Cohesion The bonding or attraction between particles of certain fine-grained soils that enhances shear strength and is independent of confining pressure.

Cold Joint Surface where concrete placement was interrupted, then later resumed.

Continuous Flight Auger A string of helical augers and a cutting head, used to bore a hole in the earth, into which a pile section may be set, concrete cast in place, or a tieback grouted.

Core Barrel   A cylindrical rock-drilling tool, designed to cut an annular space around a central cylindrical core of rock, which can then be removed to deepen the hole.

CPT (Cone Penetration Test)   An exploratory test for soil or unsound rock; ASTM Method D 3441-79, "Method for Deep Quasi-static Cone and Friction-cone Penetration Tests of Soil."

Crowd   The downward thrust of kelly bar or auger caused by mechanical reactions against the weight of the boring rig. It is available in some drilling rigs, for use when boring becomes difficult in hard formations.

Cuttings   Particles of soil or rock brought up by a rotary drill. See also **Spoil**.

Deep Foundations Institute (DFI))   P.O. Box 359, Springfield, NJ. (The source of many of the terms in this Glossary—for which the authors thank them.)

"Dental" Work   Work with hand tools in a pier hole, to break up and remove boulders or rock pinnacles; to remove intruding ledge; to level sloping bedrock.

Dewatering   (1) The removal of water from a construction area, as by pumping from an excavation or location where water covers the planned working surface. (2) Lowering of the groundwater table in order to obtain a "dry" area in the vicinity of an excavation which would otherwise extend below water.

Downdrag   A downward force exerted on a pier, pile, or other structural element by settling soil. Sometimes called "negative skin friction."

Drawdown   Lowering of the level of groundwater; for example, when a work area is dewatered for construction.

Drilled Pier, Drilled Shaft   A reinforced or unreinforced concrete foundation element, formed by drilling a hole in the earth and filling it with concrete. Also called a "caisson," or a "large-diameter bored pile."

Drilling Bucket   A closed rotary boring tool with its cutting edge at its base. Spoil is removed from the bucket by lifting it out, swinging it to one side of the bore, and releasing the hinged bottom of the bucket.

Drilling Mud, Mud, or Slurry   A fluid mixture of water and clayey soil, or commercial "driller's mud" which may be wholly or partly of the mineral sodium montmorillonite (common name bentonite) or other clay mineral.

Elastic Movement   Movement under load which is recoverable when the load is removed.

"Elephant's Trunk"   A collapsible conduit of fabric or plastic which, when coupled to the bottom of a concrete hopper, directs the concrete to a point below the reinforcing cage to prevent its striking the cage or the sides of the shaft; not used for placing concrete under water.

Expansive Soils (also **Active Clay**) Soils which shrink on drying, and expand on wetting (sometimes with great force). Under special circumstances some soils may become expansive due to heating, oxidation, precipitation of crystals from groundwater, and other causes.

Extractor A device for pulling piles or casings out of the ground. It may be an inverted steam or air hammer with yoke so equipped as to transmit upward blows to the pile body, or a specially built extractor utilizing this principle. Vibratory hammers/extractors may be especially effective.

Fill (1) Earth placed in an excavation or other area to raise the surface elevation. Fill soils are usually selected for low compressibility and good stability. (2) Controlled or compacted fill refers to material which is placed and compacted in layers under carefully controlled conditions to achieve a uniform and dense soil mass which is capable of supporting structural loading.

Fine-Grained Refers to silt- and clay-sized particles which exist in a soil.

Fixed-Head Pier A pier whose top, when deflected laterally with application of lateral force, is so restrained that the pier axis at the top must remain vertical during such movement.

Friction/End-bearing Pier A pier that achieves support from the combination of side friction and tip (end) bearing.

Friction Pier A pier which derives its resistance to load by the friction or bond developed between the side surface of the pier and the soil or rock through which it is placed.

Full-Scale Load Test A load test made on a full-scale pier or other structural element, with the load carried at least to the structural design load, and preferably to twice (or more) the design load.

Geotechnical Engineer An engineer with specialized training and knowledge of structural behavior of soil and rocks, employed to do soil investigation, to do design of structure foundations, and to provide field observation of foundation investigation and foundation construction.

Gow Caisson A short cylinder of steel plate large enough for a man to work inside; as excavation proceeds, successively smaller cylinders are set inside until the bearing stratum is reached. The telescoped cylinders are withdrawn as concrete is placed.

Gravel Small stones or fragments of stone or very small pebbles larger than the particles of sand, but often mixed with them. Generally ⅛ to 6 in. in size. (Stones 3 to 6 in. are usually called "cobbles.")

Ground Loss Subsidence of surface of ground adjacent or close to a shaft excavation, caused by soil moving into the excavation laterally during drilling, or during dewatering after drilling is complete. Common in soft organic soils or clays, and cohesionless soils below the water table.

Groundwater Level The subsurface elevation at which free water is present. (Also called "water table.")

Grout   A fluid (pumpable) suspension of cement in water, or a pumpable fluid or solution of chemicals which sets up (hardens) after a time has passed.

Grouting   Grout is pressure-injected into deposits of rock containing fissures, cavities, seams, and so on, or into permeable soil deposits, to solidify and strengthen the formation; or to reduce or eliminate a flow of water through the formation.

Gumbo   A dark-colored, very sticky, plastic clay, occurring abundantly in the central and southern parts of the United States and the prairie provinces of Western Canada.

Hardpan   (1) A dense heterogeneous mass of clay, sand, and gravel, of glacial drift origin. (2) A hard stratum of consolidated or cemented earth underlying surface soil, too hard for roots to penetrate readily.

Head   Shortened form of the phrase "pressure head," referring to the pressure resulting from a column of water or elevated supply of water.

Heave   (1) The uplifting of earth between or near piles, caused by the displacement of soil by pile driving. (2) The uplift of a previously driven pile caused by the driving of an adjacent pile. (3) The upward movement of soil and/or foundations supported on soil, caused by expansion occurring in the soil as a result of such factors as freezing, swelling due to increased water content or oxidation of sulfite soils exposed to air, or to release of overburden pressures.

Hollow-Stem Auger   An earth auger with an end bit on a hollow center shaft.

Impervious   Impervious soil is soil in which the spacing of the soil particles is so close as to allow only very slow passage of water. For example, movement of water through a typical clay (an "impervious" soil) may be only 1/1,000,000 as fast as through a typical sand.

Inclinometer   Instrument which measures the deviation from the vertical of a casing or an uncased open hole.

In situ   In place in the earth.

Kelly bar (or kelly)   A square or splined shaft which can slide vertically through a square or splined opening in a rotary driving head to turn an auger or drill bit; for a drilling rig which drives from the top, the kelly bar may be a smooth cylinder.

Kentledge   Term used by the British for weights for test loading. Also called caisson weights or load-test weights.

Kip   A force unit equal to 1000 lb.

Lagging   Wooden supports placed to support the upper surface of a large "bell" or underream in soil; used where the bell is hand-excavated. Also used to designate the planking placed between soldier piles in a shored or braced excavation.

Laitance   A fluid mixture of water, cement, and fine sand that appears at the top of concrete soon after pouring.

Large-Diameter Pile   A pile with a nominal diameter exceeding 24 in. (600 mm).

Large-Diameter Bored Pile (British usage)   Same as **Drilled Pier**, **Drilled Shaft**.

Lateral Load   That portion of load that is horizontal, or at 90° to the axis of a pier or pile, or of the supported structure.

Liner   A metal pipe, used as a temporary form when placing concrete in a pier hole; often made of 16-gauge, galvanized double-riveted corrugated metal pipe ("CMP").

Load Cell   A device for measuring the pressure exerted between the soil (or rock) and a structural element (e.g., the bottom or side of a pier); used with a hydraulic or electrical indicating or recording instrument at ground surface.

Mud   See **Drilling Mud**.

Mud Pit   A shallow pit, excavated adjacent to a boring location, used to contain drilling mud (slurry) during drilling.

Mudding-In   The technique of stirring soil and water by an auger, sometimes with the addition of commercial "driller's mud," to form a slurry as the hole is advanced by auger drilling.

Multiple Underreams   Additional underreams cut in a bearing soil, at elevations above the bottom underream, to force shearing resistance in the soil into a larger peripheral surface.

Natural Moisture Content   Moisture content in situ, at the time of measurement or investigation. May be subject to seasonal variation.

Negative Skin Friction   Effect of settling soil that grips a pile or pier by friction and adds its weight to the structure load. Also called **Downdrag** (a better term).

NX Core   Rock core taken with an "NX" core barrel, which cuts a core 2⅜ in (60 mm) in diameter.

Organic Soil   Soil containing an appreciable content of decayed vegetable matter; sometimes fine-grained, semiplastic, sometimes fibrous (e.g., peat); usually gray to black in color.

Penetrometer   An instrument for measuring the resistance to penetration of a soil to a point of defined size and shape (see **CPT**).

Permeability   The ability of a soil to permit water (or other fluid) to flow through it by traveling through the void spaces. A high permeability indicates flow will occur rapidly, and vice versa.

Pier   A columnlike foundation element similar to a pile. The pier is generally considered a foundation which is constructed by placing concrete in a deep excavation large enough to permit visual inspection.

Pinnacled Rock   Rock extending upward into the line of a pier excavation, but not filling the entire area of the boring.

Plasticity   Term applied to fine-grained soils (such as clays) which when moist can be remolded without raveling or breaking apart.

Plate Load Test   A load test made on a steel plate, resting on a surface, as at the bottom of a pier hole.

Pore Pressure   Water pressure developed in the voids of a soil mass. Excess pore pressure refers to pressure greater than the normal hydrostatic pressure expected as a result of position below the water table. Note that pore pressure may be negative, that is, tension, as established by capillary forces.

Pressuremeter   An instrument for in-situ testing of mechanical properties of a soil or rock by hydraulically expanding a probe in a bore hole, and measuring the volume changes produced by successive increments of pressure.

Proof-Testing of Rock   A test of the soundness of rock at a pier bottom, performed by drilling a hole and probing the rock in the walls of the hole by use of a chisel-pointed probe bar, or by other means.

Raise Bit   A bit for enlarging a pilot hole ending in a cavity, cavern, or mine working, by drilling from the bottom up instead of from the top down.

Rat Hole   An oil field term. A rat hole driller moves onto the oil-well site in advance of the well rig, and drills a 40-ft hole to start the well. Makers of rat hole drilling tools make rock augers for the foundation drilling industry also.

Rebar   A bar of reinforcing steel.

Retaining Wall   A structure constructed to withstand the lateral pressure of earth behind it and its own weight imposed on the soil beneath it.

Reverse Circulation   A counterflow method of circulating drilling fluid and spoil in a drill hole. In the direct circulation method drilling fluid is pumped down a hollow drill pipe, through the drill bit, and back to the surface in the annular space around the drill pipe; and the cuttings are carried to the surface by the flow. In the reverse-circulation or counterflow system, drilling fluid is pumped out of the drill stem at the top, circulated through a pit where cuttings are removed, and returned to the annular space around the drill stem. Circulation is upward inside the drill stem and downward outside it.

Rig, Drilling Rig   A machine for drilling holes in earth or rock.

"Rigid" Pier Action   The behavior of a pier that is so stiff (inelastic) in comparison with its surrounding material that the *distortion* of the pier itself, under load, exerts negligible influence on stress distribution in the surrounding soil or rock.

Rock A naturally occurring mineral substance cohesively bound by chemical bonds and forming the basic structure of the earth's crust.

Rock Auger An auger-type drilling tool, equipped with hard-metal teeth to enable it to drill in soft or weathered rock, hardpan, and so on, where a simple soil cutting blade will not penetrate.

Rock Socket That portion of a shaft which penetrates into a rock formation beneath less competent overburden.

Rotary Boring A method of boring using rotary (as opposed to percussive) means of excavation.

Rotary Drill Rig A rotary drilling machine powered hydraulically, pneumatically, electrically, or mechanically to bore exploratory holes or for installation of piers, caissons, or in-situ piles. The equipment may use a continuous-flight auger or a rotary table and kelly bar with various attachments and tools to perform the work. See **Continuous Flight Auger** and **Kelly Bar**.

Sand Cohesionless soil whose particle sizes range between about 0.05 and 2 mm in diameter.

Seepage Small quantity of water percolating through a soil deposit or soil structure.

Segregation Separation of poured concrete into zones of coarse aggregate without fines, and sand–water–cement without coarse aggregate.

Settlement (1) The amount of downward movement of the foundation of a structure or a part of a structure, under conditions of applied loading. (2) The downward vertical movement experienced by structures or soil surface as the underlying supporting earth compresses.

Sidewall Grooving The cutting of circular or spiral grooves in the walls of a pier hole in rock or soil, with the objective of improving the sidewall support of the pier.

Sidewall Shear Frictional resistance to axial movement of a pier or pile, developed between the soil surrounding the pier and the peripheral surface of the pier. (Does not include resistance to movement of an enlarged base, due to development of shearing strains within the soil below the base.)

Silt A fine-grained nonplastic soil; often mistaken for clay, but quite different in its behavior.

Skin Friction Resistance to shearing motion between the concrete of the shaft and the soil or rock in contact with it; may be actually friction between the two bodies, or shearing strength of cohesive soil retaining the pier, or shearing strength of the bond between concrete and rock socket.

Slump A measure of consistency of fresh concrete. The test is made with a truncated cone 12 in. high, 4 in. in diameter at the top, and 8 in. in diameter at the bottom, filled with concrete in three lifts, each lift rodded thoroughly 26 times. The cone is then lifted smartly and the concrete, no longer supported, subsides into a slump condition. The height of the slumped concrete in inches deducted from the 12-in. cone height is the slump expressed in inches.

Slurry   See **Drilling Mud**.

Slurry Trench   A narrow trench with vertical unbraced walls, in which caving or sloughing of the earth walls is prevented by the hydrostatic pressure of the "slurry" or "mud" with which the trench is filled. Excavation of the trench is performed through the slurry, the trench being kept filled with added slurry as excavation proceeds.

Soil Analysis   An investigation of the earth in the area of a foundation consisting of sampling, classification, preparation of logs of borings, and a report setting forth conclusions and recommendations. It is basic to the design of foundations and is required by most up-to-date building codes.

Soil Stabilization   Treatment of soil to improve its properties; includes the mixing of additives and other means of alteration such as compaction or drainage.

Sonotube   A cylindrical form of treated cardboard, for forming round columns of concrete; a commercial product.

Spiles   See **Lagging**.

Spoil   Soil or rock removed from an excavation; to be wasted or used elsewhere as fill.

SPT   See **Standard Penetration Test**.

Squeezing Ground   A soil formation, usually of clay, silt, or organic material, which tends to bulge or squeeze into the hole during drilling, or afterward if the hole is left uncased.

Standard Penetration Test (SPT) (N)   The number of blows required to drive a 2-in. O.D., 1⅜ in. I.D., 24-in. long, split soil-sampling spoon 1 ft with a 140-lb. weight freely falling 30 in. The count is recorded for each of three 6-in. increments. The sum of the second and third increments is taken as the N value in blows per foot. (This is ASTM Designation D 1586).

Straight-Shaft Pier   A pier (cased or uncased), poured in a drilled hole without underream or bell.

Strain   Relative movement between or within structural elements. For example, shearing strain between pier sidewall and surrounding soil; compressive strain within concrete of pier shaft.

Strain Gauge   An instrument or device for measuring relative motion (compression, elongation, shear, and so on) between two points in a mechanism or in a structural member.

Swelling Soil or Rock   A soil or rock material subject to volume increase caused by wetting, oxidation, buildup of crystals, or relaxation after load removal.

Telltale   A strain indicator, usually comprised of a sleeved freestanding rod cast in place in a drilled pier or pile to measure relative movement between the anchored (embedded) tips of two or more rods or between the rod anchor and the top of the pier or pile.

Temporary Casing   Casing left in place until concrete has been placed, or casing placed as protection for workmen or inspector.

Till   Dense glacially deposited soil formations which consist of a heterogeneous mixture of fine-grained and coarse-grained material and often contain significant quantities of boulders and cobbles.

Torque   The turning power of a shaft (or kelly); measured as the product of force times distance from the center of rotation. Expressed as pound-feet (lb-ft), (not ft-lb as mistakenly shown in many rig manufacturers' catalogs) (metric system: kilogram-force-meters or kgf-m).

Tremie   (1) (verb) To place concrete below water level through a pipe, the lower end of which is kept immersed in fresh concrete so that the rising concrete from the bottom displaces the water without washing out the cement content. (2) (noun) The hopper and drop pipe used to place the concrete underwater.

Truck Carrier   A specially built truck for mounting a drilling rig or for carrying a crane.

Twisting Bar   A tool to be attached to the kelly, used for "screwing" down casing though caving or squeezing soil, and sometimes for pulling casing.

Ultimate Load   The load on a pier or pile which results in continued inelastic movement.

Underream   Enlargement of the lower end of an augered or drilled pier hole to increase its bearing area. Also called "bell."

Underreamer, Belling Tool   See **Belling Bucket**.

Unit Weight   The weight per unit volume of a material such as soil, water, concrete, and so on. Typically expressed as pounds per cubic foot, grams per cubic centimeter, or kilograms per cubic meter.

Uplift   An upward force exerted on a pier, pile, or other structural element, by expanding soil or rock, by frost action, or by hydraulic pressure.

Vibratory Driver/Extractor   A pile-driving and extracting machine which is mechanically connected to a pile or casing and loosens it while driving or pulling by oscillating it through the soil. Power source may be either electric or hydraulic.

"Walking Off"   Tendency for a rotating bit to deflect laterally when encountering a sloping surface, boulder, or cobble, and so on.

Water Content   The ratio of the quantity (by weight) of water in a given volume of soil mass to the weight of the soil solids, typically expressed as a percentage.

Water Table   The subsurface elevation at which free water will usually be present. Also called "groundwater."

Well Point   The perforated end section of a well pipe which permits groundwater to enter the pipe for collection and disposal or use.

Index

Page numbers in **boldface** type refer to illustrations.